EDISON'S
CONCRETE PIANO

EDISON'S CONCRETE PIANO

FLYING ★ TANKS

SIX-NIPPLED SHEEP

WALK-ON-WATER SHOES

and

12 OTHER FLOPS

FROM GREAT INVENTORS

JUDY WEARING

ECW PRESS

Published by ECW Press, 2120 Queen Street East, Suite 200,
Toronto, Ontario, Canada M4E 1E2 / 416.694.3348 / info@ecwpress.com

LIBRARY AND ARCHIVES CANADA CATALOGUING IN PUBLICATION

Wearing, Judy
Edison's concrete piano : flying tanks, six-nippled
sheep, walk-on-water shoes, and 12 other flops from great
inventors / Judy Wearing.

ISBN 978-1-55022-863-2

1. Inventions—Miscellanea.
2. Inventors—Biography. I. Title.

T47.W42 2009   609.2'2   C2009-902544-2

Editor: Emily Schultz
Cover design: David Gee
Author Photo: Tom Riddolls
Text design: Tania Craan
Typesetting: Mary Bowness
Printing: Webcom   2   3   4   5

The publication of *Edison's Concrete Piano* has been generously supported by the Ontario Arts
Council, by the Government of Ontario through Ontario Book Publishing Tax Credit, by the
OMDC Book Fund, an initiative of the Ontario Media Development Corporation, and by the
Government of Canada through the Book Publishing Industry Development Program (BPIDP).

Canada   ONTARIO ARTS COUNCIL
           CONSEIL DES ARTS DE L'ONTARIO

PRINTED AND BOUND IN CANADA

ECW PRESS
ecwpress.com

for Tom, for everything

CONTENTS

THE **HISTORIC** AGE

THE **GOLDEN** AGE

THE **MODERN** ERA

ACKNOWLEDGMENTS

Like every invention, this book was sparked by an idea. In this case, the spark arose from a discussion with Scott Taylor, who introduced me to Thomas Edison's concrete piano and Walter Christie's flying tank; thank you Scott. From the beginning, Jack David has been a model of patience and professionalism, as well as going above the call of duty to teach a curious new recruit about publishing. I was also privileged to have the brilliant Emily Schultz as the midwife of this work.

Along the way, many have been generous with their expertise and time, and their contributions to this book's content are hefty. George Davis, George Davison, Danny Hillis, Kim Hunter, and Robert Lemelson took the time to speak with me by phone. Email questions were answered at length by Peter Abrahams, Lalita Bharadwaj, Terry Bridges, John Cline, Chathan Cooke, Greg Durocher, Isabel Deslauriers, George Hoffman, Paul Israel, Matthew McCarry, Robert McKinley Jr., Jeff Miller, Bob Siegel, Anne Stewart, Joseph Wearing, and Joe Wolfe. Paul Soderman was particularly helpful, combining engineering expertise with a knack for the written word. The members of the Armchair General, Mechanical Music Digest, and Yesterday's Tractor forums were generous and enthusiastic with their stories and insights.

Peggy Conrad, Alice Hargrove, Duane Hyland, Alicia Kopar, and Kelly Kost provided some images and additional information.

William Higginson killed two birds with one stone; he not only gave me particularly valuable access to research facilities, but he also proved that it pays to talk to strangers. Tom Goldsmith lent his talented hand and unique sense of humor to assist with promotional artwork, and Tom Riddolls interpreted and illustrated the inventions for which images are unavailable.

Sarah Bates, Terry Bridges, Isabel Deslauriers, Joshua Lohuis, Jessica Morrison, and Esther Robinson read small bits of the manuscript and greatly improved those bits. Tom Riddolls was close enough to be roped into critiquing the entire manuscript, and Nancy Foran was a wonderful, official editor.

Despite this assistance, I am certain to have still managed to produce a work that is less than perfect; all faults are my own.

Jacob and Tom did an excellent job of suffering through ramblings and emotional outbursts *ad nauseam* as each inventor overtook daily life and, one by one, became an obsession. A big thank you to all my other family and friends, who have been persistent in their encouragement, particularly my Gran, Elizabeth.

And finally, for many reasons that cannot be expressed in print, none of this book would have been possible without Tom, to whom I am happily indebted always.

Great Inventors and Failure

A great invention has the same magic as a great painting. Both are expressions of creativity — that most human of traits — and both leave us breathless with the staggering heights of their peaks next to our own relatively lowly position. Invention is innately interesting, arguably, perhaps, because it sums up in one word all that is modern humanity. Alan Kay, inventor of the computer's graphic interface, describes anthropological universals as all those things children learn from their environment. Everything else, including reading and writing, is an invention. Buckminster Fuller described the collective pile of inventions as "wealth." As collective an enterprise as invention is, it is also immediately individual, as author Tom Wolfe so eloquently explains: "An invention is one of those super-strokes, like discovering a platinum deposit, or a gas field, or writing a novel, through which an individual, the hungriest loner, can transform his life overnight, and light up the sky. The inventor needs only one thing, which is as free as the air: a terrific idea." Invention, then, is inherently great, no matter its source or its scope, and judgment of it is entirely and utterly subjective.

In writing this book, I was forced to delineate "great inventors" and "not-so-great inventions." I have done so relatively broadly in recognition of the subjectivity of these terms. I selected great inventors who have, or are thought to have, made a significant contribution to society. The inventions featured have failed in one respect or another, either by failing to make a significant impact on society or by exhibiting flaws that are surprising. The concept of failure is just as loose, just as subjective, as the concept of great. If I've done my job well, this subjectivity will be evident in the selected examples.

Looking at the lives and work of numerous great figures in the history of invention at once reveals patterns in the characteristics of greatness and the nature of success and failure. For example, most of the inventors discussed had markedly strong characters — the sort that did not back down easily in the face of adversity. Malcolm Gladwell's supposition that the self-made man is a myth is not well upheld by the stories of the inventors in these pages. These great inventors certainly had mothers, wives, teachers, and friends who supported them along the way, but they also had a tremendous fighting, pioneer spirit. It is this spirit that stands out as a primary factor in their achievements.

A high degree of gumption being a prerequisite for greatness makes sense, considering the inventor is inherently isolated. As Jewkes, Sawers and Stillerman point out in *The Sources of Invention*:

> His crucial characteristic is that he is isolated; because he is engrossed with ideas that he believes to be new and therefore marked out from other men, and because he must expect resistance. The world is against him, for it is normally against change and he is against the world, for he is challenging the error or the inadequacy of existing ideas. . . . Anyone who doubts the feeling of isolation that inevitably surrounds the innovator should try to put himself in the position of Leonardo da Vinci when, more than one hundred years before Galileo, he wrote in his notebook in large letters "The sun does not move."

Virtually all of the great inventors investigated for this book had recorded failed inventions, usually appended to a long list of more illustrious achievements. For certain, there are many other outright failures, or discarded prototypes, of which there is no trace. Perhaps among these discarded ideas were inventions that were rejected on the basis of perceived lack of popularity or commercial potential. And perhaps some of these rejects could have been of tremendous benefit to society later, if only they had been followed through or at least recorded. With very rare exceptions (e.g., Elihu Thomson, inventor of the three-coil dynamo, did very little "wrong" in his life), all people fail in some ways. Since we all are aware that everyone fails, why do we ignore or, even worse, disguise our failures? Homer Simpson sums it up, "If at first you don't succeed, destroy all evidence that you ever tried."

The not-so-great inventions in this book failed in the sense that they did not make a lasting impact on society (Ford's Fordson is an exception). There are a myriad of reasons for invention failure; faulty design is only one of them. Some of the inventions featured here failed because of a characteristic of the inventor. In fact, greatness in invention requires such strong vision that the greatest of inventors can be blind to the faults of their own work; to them it *is* great, no matter what anyone else has to say about it.

That said, many perfectly sound inventions are steadfastly refused by the society they are designed to aid. Time and time again there are examples of innovation, which promise to improve life, that are ignored or cast aside. Then there are the gatekeepers; the few people in charge of buying and licensing products for manufacturing and retail giants do not necessarily have a mandate for technological change or the bettering of society, nor the capacity to envision an entirely new way of doing things. Add to this the might of the dollar, the corporations that control it, and the stockholders who require a decent chance of return on investment. Some simple and effective ideas never come to market because they cannot be sold for enough to reap the necessary profits required by manufacturers.

Individual consumers can also be responsible for an invention's failure. Consumers are generally a conservative lot who look at change of any kind as a threat to personal security. The history of the loom, as told by Joseph Rossman, presents an extreme case study:

Kay, the inventor of the flyshuttle, was mobbed when he introduced his invention in Lancashire. A mob destroyed the spinning-frame invented by Hargreaves. Crampton had to hide his spinning-mule, fearing a similar fate. Jacquard barely escaped being drowned in the

Rhone by angry weavers on account of his new loom. There were
riots in Nottingham on account of the introduction of the stocking-
loom.

The farthest reaching of the inventors, those who are so far
ahead of their time that they seem (in some cases literally) to come
from another planet, are most likely to be scoffed at. It is these
inventors — the Nikola Teslas and the Buckminster Fullers —
whom society might do well to pay extremely close and serious
attention to, rather than laugh away. For these visionaries are
inventing based on a view of the world that is on a different plane
than the rest of us, a plane that time has proven to be valid in significant ways.

Great inventors had characteristics that allowed them, in varying degrees, to overcome societal resistance, including financial
acumen or sales and marketing ability. Some, like James Watt, had
others to achieve these tasks for them. Inventors and other innovators take note: if you are lacking in all of the skills required, find
others who can pull up your slack. But make sure that they are good
at what they do, or you'll end up treading water like George
Washington Carver.

Failed inventions hold lessons for society, too. For one thing, the
fertile minds of these inventive giants have developed many an idea
that has not been brought to fruition. Some of these ideas are still
feasible, still accessible in the patent banks, and still have potential
to have a positive effect on human life — or to make someone a lot
of money. Inventors' notebooks and correspondence and patent
records are great places to mine for improvements of all kinds.

The tales of independent inventors and their struggles speak of
discouragement, and worse, from the very society they are trying to
help. It is a very rare independent inventor who becomes wealthy
from his or her inventions, and only about 20 percent of these innovators turn any profit. Even Thomas Edison claims that he did not
profit from his inventions when he took his court costs into consideration; the money to be made was in their manufacture. Despite

this, independent inventors are uniquely situated to dream up major advances in technology, being unbounded by corporate mandates and limitations in process and product breadth. History tells us that the creative freedom of independent inventors pays big dividends for society — including major corporations. And yet, society does little to financially support these key individuals in order to directly reap the benefits of their work. In fact, the 200-year-old basis of patent law, designed to protect inventors, in some ways works against them in modern times. For example, in many countries today the first to file a patent secures the rights to an invention. The United States is about to change its patent law to this system; until now, independent inventors who develop an invention, build a prototype, and can prove they conceived the invention first own the rights to it. This switch from "first to invent" to "first to file" is generally recognized as favoring large corporations that have more resources to file a patent application, or numerous patent applications, to protect their ideas.

Once inventors bring forward their ideas, we, as a body of people, generally resist them. How often do we keep an open mind with respect to the proposition of something entirely new? Not often enough. It is fair to say that at any one time, technology is much farther advanced than society — it is not inventors who pull us forward, it is us who hold them back.

Thinking about what makes someone great and what makes an idea fail can provide us with ammunition for our own life battles. The inventors in this book had to face failure. The most successful of them were able to accept this failure with tremendous grace: Elihu Thomson, Buckminster Fuller, and Thomas Edison are exemplary in this regard. They seem able to stare in the face of the lack of vision of those around them, or any other setback, and trudge on undeterred, with pleasantness even, toward their bigger goals. At age 67, Edison watched as his buildings, including factories and laboratories, went up in flames. While directing the firefighting on site, Edison pulled out his notebook and began making plans for a new factory. Rebuilding began the next day. The new plans included a large water reservoir to help prevent any future fires from wreaking the same tremendous damage.

Often enough, how our society defines success (e.g., accumulation of wealth and/or accolades) may not be how we ourselves define success. This can set inventors, and the rest of us, up for failure in two ways. On the one hand, if we chase after societal success, we might be barking up the wrong tree, unwittingly. On the other, if we chase after some personal, unrealistic idea of success, we might become frustrated without due cause. The story of Leonardo da Vinci reflects this conflict between personal and societal definitions of success.

Seeking success is so infused in our society that we rarely stop to contemplate failure, and we miss out on valuable lessons as a consequence. Henry Petroski has examined the role of failure in engineering, and the failure of engineering to adequately give failure an overt role in the design process. He argues fervently for its conscious inclusion in the field. "The concept of failure is central to the design process, and it is by thinking in terms of obviating failure that successful designs are achieved. . . . Indeed, the history of engineering is full of examples of dramatic failures that were once considered confident extrapolations of successful designs; it was the failures that ultimately revealed the latent flaws in design logic. . . . Design studies that concentrate only on how successful designs are produced can thus miss some fundamental aspects of the design process . . . yet while practicing designers especially are notorious for saying little, if not for consciously avoiding any discussion at all of their own methodology." The role of failure in invention is hardly surprising, considering that the nature of invention itself is more art than reason, more imagination than experiment.

Generally speaking, there is much to learn in all failure, and yet our instinct, as individuals and collectives, is to hide our missed marks. By being brave enough to openly expose and discuss our shortcomings, we do not denigrate ourselves but rather raise up others.

Hopefully, given the chance, the inventors in the pages to follow would agree with me.

THE
HISTORIC
AGE

UNDERSTAND PHYSICS

LEONARDO DA VINCI's
WALK-ON-WATER SHOES •
1452–1519

This drawing of a soldier walking on water is found in the Codex Atlanticus, *one of da Vinci's notebooks.*

From obscure beginnings in a small town in Italy over 500 years ago, Leonardo da Vinci has become one of the most famous individuals in the history of Western civilization. His great work, the *Mona Lisa*, is the single most visited painting in the world; it also holds the record for the painting most subjected to vandalism. Whether fan or critic of this enigmatic figure, da Vinci's power to affect people's emotions is undeniable. His first biographer, the Italian Giorgio Vasari, who was eight years old at the time of da Vinci's death, wrote of him, "celestial influences may shower extraordinary gifts on certain human beings, which is an effect of nature; but there is something supernatural in the accumulation in one individual of so much beauty, grace, and might." The list of disciplines that have some claim to assess his accomplishments is long and includes art, geology, optics, anatomy, music, mathematics, botany, mechanics, general physics, astronomy, literature, theater, geography and cartography, graphics, engineering, architecture, hydraulics, and chemistry. And yet, despite his universally accepted genius and his timeless influence, da Vinci's life was full of failure at every turn. He is a paradox, as enshrouded in mystery and as intriguing by nature as Mona Lisa herself.

Da Vinci was a *homo universalis* and presents a near perfect distilment of the quattrocento.[1] As a boy he was a classmate of Botticelli, and later he rubbed shoulders with Michelangelo in Florence, as well as many other artistic geniuses who sought to understand nature as a whole. He was also an illegitimate child, a product of a brief union between a peasant woman and a notary. Da Vinci's illegitimacy had a profound influence on his failures and successes throughout his life. He was taken from his mother's household and raised by his grandparents on their small farm, though it is speculated that he likely visited his mother and her subsequent

---

1 Referring to 15th century Italy, a fertile period in cultural history and the setting for the Italian Renaissance.

family from time to time.[2] The disrupted bonding between mother
and child undoubtedly had an influence on his psyche in negative
ways that could have contributed to his failures; Freud would agree.

More directly, being illegitimate restricted his education and
choice of career. He was not allowed to learn Latin formally or go to
university, nor could he follow in his father's footsteps and become a
notary, doctor, or any of the "noble professions." Instead, the best
careers available to him were in the "mechanical arts," in many
respects equivalent to the "trades" of today. His lack of Latin certainly
impeded his book learning, and his own words suggest it impeded his
societal standing as well: "I well know that, not being a literary man,
certain presumptuous persons will think that they may reasonably
deride me with the allegation that I am a man without letters."

As a young teenager, da Vinci entered the studio of the craftsman Verrocchi, in Florence, as an apprentice. Fortunately for him,
the time and place dictated that painting, sculpting, architecture,
metalworking, and engineering were all considered to be of the
same realm, and it was normal for one man to pursue them all,
which was just how da Vinci liked it.

It is impossible to give a comprehensive list of da Vinci's successes
and failures as an inventor for several reasons. For one, the collection
of his surviving personal notebooks is incomplete. Begun when he
was about 30, the notebooks record many of his thoughts (though
presumably not all as the notes are relatively devoid of emotion).
Here we find everything from grocery lists, library catalogs, studies for
his famous artworks, anatomical and scientific studies, sketches of
inventions, explorations of physical phenomena (such as motion on
an inclined plane), attempts to solve major geometric puzzles of the
day, stories, poems, and quips and doodles of all kinds. We do not
have anywhere near the entire collection of his notes, and we do

---

2 Leonardo's father, Ser Piero, may have helped his mother find a husband, and she was married relatively quickly to an acquaintance of the da Vinci's. His father married a woman of higher standing
within months of Leonardo's birth.

not have a clear frame of reference to determine how much of the notes are copies da Vinci made from other books, improvements he made on existing material, or original concepts. We also do not know the extent to which he tried the inventions he illustrated. Da Vinci made much of "experimentation," stating, "Before you base a law on this case test it two or three times and see whether the tests produce the same effects," but no records of measurements exist.

We do know that he had a workshop studio at the Sforza court of Milan and technicians in his employ. We also know that part of his salaried duties in Milan, and later in France, included the design and building of devices used in special events and festivals that are not described on paper, such as a robotic lion. All we have are secondhand accounts of this walking mechanical beast, which was created to entertain the French King Francis I. Sixteenth-century historian Piero Parenti wrote in 1509:

> When the King entered Milan, besides the other entertainments, Lionardo da Vinci, the famous painter and our Florentine, devised the following intervention: he represented a lion above the gate, which, lying down, got onto its feet when the King came in, and with its paw opened up its chest and pulled out blue balls full of gold lilies, which he threw and strewed about on the ground. Afterwards he pulled out his heart and, pressing it, more gold lilies came out, showing how the Florentine Marzocco, represented by such an animal, had his guts full of lilies. Stopping beside this spectacle, [the King] liked it and took much pleasure in it.

The robotic lion made an impression. . . . And yet, it is absent from da Vinci's surviving notebooks. Perhaps the great man did not record all of the work he did while on the job; it is conceivable that da Vinci's notebooks record some data about his inventions but not all. All of which leads to the unfortunate conclusion that we do not have, and never will have, enough information to judge da Vinci as an inventor. Any discussion of da Vinci the inventor must take this into account, and we must assess the man and his work as holistically as possible in light of the imperfect picture we have.

There is one realm where da Vinci's success as an inventor is undisputed, and that is his artistry. His works in painting and drawing are not only among the most revered in all of European art, but they represent inventions on a multitude of levels; with virtually every painting, da Vinci turned convention on its head. At a minimum, he heavily influenced his peers and future schools of artists; in some cases, he single-handedly advanced an art form. For example, his *Study of a Tuscan Landscape* is the first dated landscape study in the history of Western art. He introduced a preparatory sketch style that featured alternate lines, and the systems of light and shade he created were novel among Italian painters. His use of tone, color, reflection, and shine are unique. Raphael and Giorgione both openly copied da Vinci's masterful style. He played a role in influencing portraiture by creating a personal connection between the viewer and the subject, and his works were key to the birth of Mannerism. He invented compositional motifs, such as the pyramidal arrangement of the Virgin with Christ and St. John, as seen in the painting *Madonna of the Rocks*, and the placement of Christ on the same side of the table as his disciples in *The Last Supper* (something that would have been considered very radical in da Vinci's day). These new forms of iconography were repeated by others for generations. His techniques were also unique; he applied multiple layers of paint so thinly that X-ray imaging reveals only ghostly images instead of the sharp outlines revealed in other artists' work. Da Vinci was a master of technique, as well as form, light and shadow, atmosphere, composition, viewer psychology, motion, gesture, and drama.

Da Vinci's art was a scientific and engineering endeavor. He took great pains to study and understand light, movement, anatomy, and any other branch of science that had an impact on his art. His understanding of light and shadow is in no small part responsible for the magical qualities of his paintings. His understanding of perspective and perception allowed him to manipulate reality in scenes such as *The Last Supper*, making us party to his suspension of natural dimension, light, and space in order to experience his desired effect — a personal encounter with the 12 disciples seated for dinner at a dramatic moment. His study of anatomy was exemplary, not only for its artistry but also its accuracy. Though not infallible (for example, he connected "the spinal chord to the penis in order to transmit the vital spirits into the sperm"), da Vinci's observation skills were astute. His detailed description of the form and function of the heart's aortic valve and the flow of blood within it have proven to be accurate in recent decades through the use of imaging technology.[3]

Da Vinci's anatomical drawings represent another facet of his invention — the techniques of scientific illustration. Virtually every aspect and graphic effect used up to modern times to communicate anatomical form and function through illustration can be found in these drawings, including 3-D shapes within a transparent body, sections, inset magnifications of key features, rotation of solid forms, functional diagrams, and transparent layers.

Despite all the success of da Vinci's artistry, he also had many failures. For one, he did not finish many paintings, relatively speaking. He sold few paintings and no sculptures. Many of his most famous works were never passed on to the people who paid him to paint them. He failed to get jobs he wanted, abandoned commissions he was given, and left contractual obligations unfulfilled. He experimented with important works by using new techniques, sometimes with disastrous results. For example, a battle scene to be painted on a wall in Florence, for which he expended great efforts in planning and drawing, was done with a new blend of paint, which ran down the wall in a mess before it could dry. Likewise, *The Last Supper* was painted with experimental techniques and deteriorated quicker than it should have as a result. A similar story of impractical plans that came to naught can be found in a statue

3 Da Vinci critic Clifford Truesdell points out that his study of the heart was restricted in general to its structure, and that he did not take further steps to elucidate the function of what he saw. This negation of achievement based on the absence of all possibilities seems pompous, akin to judging a scientific genius today because of a failure to answer all the great questions that could be posed.

commissioned by his longtime patron, Ludovico Sforza, the ruler of Milan. Da Vinci worked over six years, off and on, on the bronze statue that ambitiously attempted to create a larger-than-life horse and rider balanced on two feet. The scheme involved great feats of practical engineering never before attempted, which required him to invent new methods for casting, new furnace designs, and alloys. Ultimately, these ideas proved to be impractical.

Though painting may be considered the realm of his greatest success, it seems that da Vinci did not particularly like doing it. At age 30, to facilitate his aim of leaving Florence for greener pastures, he sought the patronage of Ludovico Sforza. In the letter he sent to sell himself as a potential employee, painting is very low down on his list of abilities. Instead, da Vinci talks about contributions he can make in civil and military engineering and describes inventions to aid in both. In the letter, nine items address military inventions, and the tenth states: "In time of peace, I believe that I can compete with anyone in architecture, and in the construction of both public and private monuments, and in the building of canals. I am able to execute statues in marble, bronze, and clay; in painting I can do as well as anyone else . . ." Leonardo da Vinci considered himself an inventor first and foremost, with a keen interest in military applications. Though he apprenticed with a painter, the decorative arts were not his first choice of career, and creating art, in the purest sense, was not a priority for him. People do not necessarily like doing best what they are best at doing.

The strong interest in war da Vinci expresses in his letter to Ludovico Sforza is ironic, for da Vinci was a peaceful sort. He called war "a most beastly madness." This was a man who held life sacred to the utmost. He was a vegetarian and is reported to have walked through the markets buying birds for the sole purpose of giving them freedom. At first blush, it seems deeply hypocritical for him to have pursued military engineering as a career, so to speak. However, in his time, military engineering was the most illustrious of the mechanical arts, the highest post attainable to him, and the one that was the best paid.

The status accorded to military engineers was a practical consequence of the importance of defense and offensive capacity of the leading faction, for 15th- and 16th-century Italy was a place of uncertain politics and certain conflict. Da Vinci seems to have been good at playing the game, managing to keep himself in employ through allegiance to the right men at the right time and moving when necessary to avoid any nasty consequences, such as hanging and quartering. His military inventions were profuse, and he was in high demand. In his notebooks, we find fortified walls, explosive shells, ballistas,[4] arquebuses,[5] assault vehicles, and a mobile lock to flood a river at will and drown an army, Red Sea style.

The designs for these devices are impressive, though the extent to which da Vinci had a hand in inventing them, or attempted to build them, is unknown. All that can be said with confidence is that da Vinci was innovative in his approach to military engineering. Not all of his designs were practical however. For example, his giant crossbow, a device he returned to many times, defied physics and couldn't work. Another example of his impractical invention is an outrageous design for a shield; the drawing shows a shield fitted with a trapdoor that opens to seize an approaching sword. Da Vinci seems to have gotten his fill of military deployment when he joined the entourage of the murderous Cesare Borgia, Machiavelli's inspiration for *The Prince*, for less than seven months in the winter of 1502. After this, his interest in military engineering waned.

Da Vinci's other inventions include practical mechanical items like locks, gears, pumps, pulleys, jacks, bearings, hinges, axles, and springs, as well as larger contraptions like textile machines, mills, engravers, timekeepers, plumbing, devices to escape from prison and swim underwater, a drawing machine, a camera obscura, an automatic street washer, giant excavators, water pumps, a para-

---

4 A siege weapon that projected large objects, rather like a giant slingshot of sorts. Ballistas were used in ancient Greece and Rome.

5 A muzzle-loaded firearm used from the 15th to 17th centuries; a forerunner to the rifle.

chute, machines to make rope and coins, and even a sort of plastic. One of his later designs for a glider has been tested in modern times with success.[6] Again, it is unclear to what extent, if any, these devices and gadgets were built or used in da Vinci's lifetime.

Yet grander schemes included a town plan with three levels: the upper pedestrian level was reserved for gentlemen, the ground level bustled with the general populace and roads for the distribution of goods, and an underground level housed a sewage system with good circulation to keep disease at a minimum. He also dreamed up self-cleaning stables and a means to drain the Pontine Marshes. More plans recorded in his notebooks include: shaving off the tops of hills to facilitate a line of fire, escape tunnels, joining Florence to the sea via a canal, and building a bridge that joined Europe to Asia. One of these schemes was actually tried during his day; troops were deployed to dig a trench west of Pisa to redirect the river Arno from the sea. Unsurprisingly, it failed miserably. It was overambitious and grossly miscalculated. The river refused to flow into the channel, preferring the path of its own making.

Da Vinci's walk-on-water shoes were part of one of his more grandiose schemes — to wage war via water. The image of a man wearing the shoes appears in the Codex Atlanticus amid drawings of machines and devices to lift water. But it was part of a military line of thought that included several ideas, such as sending divers into combat armed with sharp cutlasses and drills to poke holes in the hulls of enemy vessels. While totally sci-fi in character, the concepts are not completely outrageous. "Frogmen" are employed by modern military forces for a variety of tasks today, including infiltrating enemy vessels, reconnaissance, transportation of troops, deactivation of explosives, and planting weapons in enemy territory.

From the walk-on-water shoes drawing, we can deduce that da Vinci's basic concept involved inflated attachments to the feet with poles that end in similarly inflated attachments. Bladders from some

---

6 See YouTube for footage of a modified version of da Vinci's glider in use; search for "da Vinci glider."

domestic animal were the likely material imagined for the flotation devices. We know that da Vinci was familiar with, and used, inflated offal. One of the best stories relating his relationship with this organ byproduct describes a special event at court. Ever the practical joker, da Vinci blew up some well-cleaned bullock's intestines in a crowded room. The natural balloon stretched and stretched until it squished all the guests in the room against the walls. Quite a party trick!

Note that da Vinci's walk-on-water drawing is full of personality. It is not a technical, descriptive drawing, such as one might find in a patent application, but rather the pen strokes are brimming with motion and style. With his few lines we see in full force how he envisioned walk-on-water shoes could be used. He makes it look efficient, perhaps even fun. While he certainly was perfectly capable of drawing technically in order to communicate ideas regarding form and measurement, he did not do so in this case. Perhaps this is indicative of a sketch that is more of a fleeting mental image than an actual invention — a visual artist's equivalent to thinking out loud. On the other hand, there is equivalent artistic flourish in several of da Vinci's drawings of more serious devices. For example, his design for a device to knock over the ladders of would-be wall-scalers includes a number of cute animated characters to illustrate its operation. A highly mechanical drawing of a device to lift water features the picturesque backdrop of a stream descending from the mountains and through a plain before being delivered to the foot of a water pump; it is a breathtakingly beautiful scene that brings the machine to life. There are other drawings of machines from roughly the same period; they do not include these flourishes. It is a "Leonardoism."

Walking on water is very difficult to do. So difficult that it is the stuff of gods and superhumans. Jesus is reported in the gospels[7] to have miraculously walked halfway across the Sea of Galilee,[8] a

---

7 E.g., John 6:16–21.

8 Lake Kinneret, Israel.

freshwater lake approximately 13 miles (20 km) across, to meet his disciples in their boat. He is not the only one; the Egyptian god Horus walked on water, as did Orion, a god of Greek mythology. In nature, only one animal larger than a quarter has accomplished the feat, and even then it does not do it very well. The basilisk lizard of New World rain forests, otherwise known as the Jesus Christ lizard, escapes predators by dropping into water and running across the surface for up to 15 feet (4.5 m), at a speed of about 5 feet per second (1.5 m/s), before trading running for swimming.

The obstacles to walking on water are the same for humans and reptiles. First, you need buoyancy in order to stay afloat. Da Vinci's inflated bladders might have accomplished this. Second, you need to be able to balance, which is particularly tough for humans because our center of gravity is relatively high off the ground. Rounded as they are, inflated bladders underfoot, as depicted in da Vinci's drawing, would make it very difficult to remain upright. Something flat, like a piece of Styrofoam, is much more likely to do the trick. The use of poles might help, and it is to da Vinci's credit that he realized this and provided his imaginary water-walker with them. However, another obstacle to be overcome might also make use of the poles — propulsion. While the muscles in our legs can provide the force required to move forward, they do so by pushing against something. Water does not provide the same resistance that hard ground does. When a would-be water-walker pushes against water, it moves out of the way. Even if a person can float on water *and* remain upright, moving forward is very tricky. So tricky that da Vinci may have done well to heed the African proverb "only a fool tests the depth of the water with both feet."

However, attempting to walk on water is not as far out as it first sounds. There have been over 100 patents for walk-on-water devices given by the United States Patent Office, including several for "inflatable water skies for individual use," reminiscent of da Vinci's 500-year-old concept. The most recent inflatable water ski, patented in 2008, features an inflatable paddle to "facilitate transport and reduce required storage space."[9]

More surprisingly, walking on water has been accomplished to no small degree in recent decades. Rémy Bricka of France has walked across the Atlantic Ocean! Over the course of 40 days in 1988, Bricka traversed the 3,500 miles (5,636 km) from Tenerife to Trinidad. He did so with polyester skis, dragging a catamaran for a bed and a water filter. He brought no food and survived off plankton, the occasional flying fish, and the 60 pounds of fat and muscle he shed during the journey. Bricka also holds the record for the fastest walk-on-water stint. In 1989, he walked one kilometer in just over seven minutes in an Olympic-sized pool in Montreal, Canada. No one has covered the same distance in less time since. Walking on water is Bricka's hobby; his real job is playing in a one-man orchestra, with symbols between his calves and a white dove (or rabbit) sitting on his shoulder. If you're interested, he's got a website and a MySpace page with videos and recordings.

Walking on water is also the goal of the engineering department at the University of San Diego, which hosts groups of local high school students each year for a walk-on-water competition. Similar engineering skill competitions are held sporadically around the world. The object is always the same: teams compete to design and build walk-on-water devices that allow a person to walk on water in a pool. Assistant Professor of Mechanical Engineering Matthew McCarry runs the USD event. In an email he wrote, "humans tend to fall over when standing upright on floating objects. (Don't believe me? Watch people who are learning to surf.) When standing on a floating surface we also need a way for the surface to 'dig' (propulsion) into the water, otherwise you walk in place."

McCarry says that da Vinci's shoes "are too short and narrow and don't enable the user to have the center of gravity directly over the shoes at all times. Without the poles the center of gravity would shift rapidly from side to side and cause the user to topple over. Also, it is unclear from the picture how propulsion would be accom-

---

9 US Patent Number 7,361,071.

plished." In the modern competition, well-designed shoes are "lightweight for buoyancy, long and wide to keep the center of gravity above the shoes, and have flaps to aid in propulsion." While a lot of fun, McCarry sees no practical application for the invention, as they would be so difficult to use, except perhaps to train athletes.

Da Vinci's walk-on-water shoes were a failure. There is no record of Renaissance Italian soldiers terrifying enemy sailors with miraculous feats on water and no evidence of floating shoes. The concept may have had some merit, but the design of bladder shoes and poles is impossible to the point of ridiculous. Brilliant da Vinci, whose chief preoccupation in life was to understand the fundamental laws of the universe, made a big error of judgment, despite his profound accomplishments in understanding the physics of water. Just as da Vinci believed that in order to understand he needed to consider everything, so we, too, cannot judge the failure of this one invention in isolation. Instead, it must be considered in the context of a single drawing among many that depict innovations to wage war on water and found in volumes of private notebooks that contain hundreds upon hundreds of drawings of innovations, which represent only about 20 percent of the estimated total notebook material da Vinci produced, which still does not represent the extent of the ideas he realized, let alone imagined. It is a doodle, a thought, a whim. And a rather imaginative one at that.

There are, however, bigger da Vinci failures to address. He failed in a multitude of ways throughout his life — including making errors in describing physical laws, the development of ideas that he truly believed were possible despite their complete impracticality, and many grand schemes that came to naught. Generally speaking, his driving desire to understand the world was frustrated, and his goal of communicating all he had learned remained an elusive dream. In fact, his notebooks were uncaringly dispersed years after his death and were not made readily available to the public until Victorian times.

Throughout the notebooks we do have, da Vinci repeatedly

expressed a sense of his failure. He penned the phrase "tell me if anything was ever done" numerous times. A prominent da Vinci scholar believes this is a "verbal doodle . . . the phrase which sprang most readily to mind when he had to try a new pen, or during moments of abstraction from the immediate business at hand."

As a young man, da Vinci declared, "I wish to work miracles; it may be that I shall possess less than other men of peaceful lives, or those who want to grow rich in a day. I may live for a long time in great poverty, as always happens and to all eternity will happen, to alchemists, the would-be creators of gold and silver, and to engineers who would have dead water stir itself into life and perpetual motion, and to those supreme fools, the necromancer and the enchanter." As he grew old, da Vinci declared that he had attempted too much and had confused himself in the bargain. He said, "I have wasted my hours" and "nature is full of infinite causes that experience has never demonstrated." The great observer and self-proclaimed experimenter had been defeated. Then, after pursuing a deeper and holistic understanding of the world for over 60 years, on his deathbed da Vinci is reported to have declared "how much I have offended God by not working on my art as much as I should have."

Many historians have not been particularly kind to da Vinci when weighing up his more scientific achievements. Vasari, his first biographer, in 1550 stated that "in learning and in the rudiments of letters he would have made great proficience, if he had not been so variable and unstable, for he set himself to learn many things, and then, after having begun them, abandoned them." Some 20th-century essays on the history of science have proclaimed da Vinci's influence on scientific thought negligible, in part because his notebooks were squirreled away, unseen, while science progressed without them. More damningly, these essayists assert that the quality of his efforts is far less than worthy of the praise commonly heaped upon him. "Leonardo in method and expression was much the same as a painter, as a mathematician, and as a physicist: In each domain, his eye and hand leave far behind his brain and spirit."

Professional jealousy aside, we can state unequivocally that da Vinci considered himself a failure, whatever the rest of the world thinks. If success is defined by reaching one's own goals, then he failed miserably in every respect, walk-on-water shoes included. However, if success is having a lasting impact on the world, then he is an astounding success — in his artistry, most certainly, and in his invention, probably (though the extent to which his innovations realized in 15th- and 16th-century Italy influenced the machinery of the day we can never know). In less concrete ways, the da Vinci presented in his notebooks has an impact on our collective psyche. He forces us to decide whether success is subjective or objective. Who is the better judge of a man's worth? Himself or others?

Invention is imagination translated into a material tool. In this light, da Vinci's walk-on-water shoes are an excellent first step, all the more so if viewed as a playful drawing of a concept rather than a blueprint. That the shoes did not get off the drawing board is less the result of lack of ability or error than the product of an incredibly fertile mind that was spread so thin and attempted so much that failure was inevitable. Perhaps failure in invention can be viewed as a simple mathematic principle, for invention is by nature the product of variation in ideas. The greater the variation, the greater the quantity of high-quality results, but also the greater the quantity of low-quality results. Da Vinci said, "like a kingdom divided, which rushes to its doom, the mind that engages in subjects of too great variety becomes confused and weakened."

The more one tries, the more one fails.

START WITH A GOOD IDEA

JAMES WATT'S APPARATUS FOR ADMINISTERING MEDICINAL AIRS

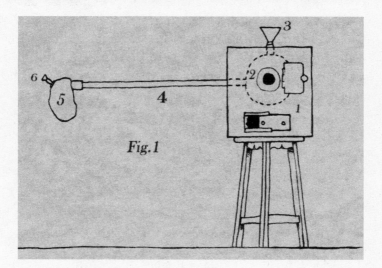

Fig. 1

This artist interpretation of Watt's apparatus features a furnace for heating up materials. The patient would inhale the airs through a bag on the end of a long tube. (See endnote.)

1736–1819

In a modest home in cold 18th-century Scotland, a young, sickly boy watched with fascination as steam rose from a kettle on the woodstove. So goes the legend of James Watt and his first thoughtful encounter with steam. As legends go, it paints a romantic picture, however imaginary. Watt's first real encounter with the power of steam came when the young-adult Watt was working as an instrument-maker on the campus of the University of Glasgow. A model of the Newcomen steam engine[1] used for demonstrations in student lectures had broken and wound up in Watt's shop for repair. This chance event set Watt on a journey with steam that would last a lifetime.

He was fascinated by the workings of the engine but mortified by its inefficiency, and he set a course to figure out how it could be made better. The Newcomen engine consisted of one cylinder and one piston. When at rest, the piston was held at the top by counter weights. The cylinder was attached to a boiler, which supplied steam to fill the cylinder. Then, the cylinder was cooled by the injection of cold water. This condensed the steam, creating a vacuum and pulling the piston down. The process was repeated by heating it all up again. This heating and cooling took tremendous amounts of energy, but Watt changed all that. His first, and biggest, contribution to the steam engine was to devise a separate condenser chamber. The cylinder could then remain hot. To create the required vacuum, a valve in the cylinder was opened, releasing the steam into the separate condensing chamber, which was filled with cold water. This created the necessary vacuum in the cylinder without reducing the temperature of the piston's cylinder. With this advancement, Watt created a workable model that could be used broadly in industry; the modern steam engine was born.

Watt did not stop there, however. Over his career, he developed other steam-engine models with various improvements, one of the

---

1 Thomas Newcomen invented this steam engine in 1712. It was the first practical machine to use steam to do mechanical work, though it was highly inefficient. Its chief use was to pump water out of mines.

most important being the rotating engine, which could turn a
wheel. All previous models drove a lever up and down, which had
limited applications, such as pumping water out of mine shafts.

Watt was an instrumental cog in the wheel of the industrial revolution, but his creative genius had broader scope. At age 43, when
he grew tired of copying out business letters by hand to send them
to more than one person, he invented a copying machine. The
machine worked by applying pressure to the writing against a damp-
ened sheet of paper. Selection of the right types of ink and paper
were important to the device's operation. A press took the copies
off the device. This form of offset printing was a tremendous success, furnishing Watt with an income while the steam engine was
in its infancy.

Watt also had a lifelong interest in the technical aspects of art
and invented a tool for drawing in perspective, as well as another
for making duplicates of sculpture or any other three-dimensional
object. The latter he worked on in his later years, and he was still
making improvements to it at the time of his death. He made contributions to civil engineering, including survey instruments and a
jointed, flexible pipe to move water from a natural spring across the
River Clyde to supply the houses of Glasgow. He was a decent scientist as well, and there is an argument to be made that he was the
first person to work out the composition of water.

Watt did have some failures, though they were few and far
between. He described the nature of failure in his profession in conjunction with a new idea thus, "I have been turning some of my
idle thoughts lately upon an arithmetical machine. I intend to make
an attempt to make it; I say an attempt, for though the machine is
exceedingly simple, yet I have learnt by experience that in mechanics many things fall out between the cup and the mouth." Another
early idea for a circular engine — "a wheel with a hollow rim which
could be filled with steam, and three steam-pipes acting as spokes"
— was also a dead end.

Much has been said about the personality of the successful inventor: their drive to create, their tenacity in the face of adversity, and

their propensity for risk. Watt does not fit this picture. His numerous biographies are peppered with evidence of a very different personality type — one that is prone to self-doubt, fear, and reluctance.
In a letter written early on in the steam engine's history, Watt
ascribed his lack of progress to his "natural inactivity, want of
health and resolution." At age 36, Watt described himself in self-
deprecating terms to an acquaintance who might employ him in
time of need, "I can on no account have anything to do with workmen, cash, or workmen's accounts. . . . Remember also I have no
great experience and am not enterprising, seldom chusing [sic] to
attempt things that are both great and new. I am not a man of reg-
ularity in Business and have bad health."

Watt's writings tell of a long-standing battle with negativism
and fear of failure, even in times when things were going well. In
1779, with his steam engines selling and business booming, he
wrote in a letter: "to my sorrow I find that as Business encreases
[sic] my Cares encrease in the same ratio which in a very considerable degree prevents the enjoyment of life." A year later he
described his circumstances thus, "our prospects are brilliant, our
income increases yearly" but then went on to say, "I am so habitu-
ated to disappointment that even these splendid prospects cannot
raise my spirits to par."

In his forties Watt described his memory as poor and associated
it with his heartsick mood, and he complained repeatedly of
"stupidity." Watt was a harsher critic than he needed to be, though
his fortunes seemed to have been buoyed at every turn by a circle of
friends who recognized his ability and business partners that were
astute at their trade, and honest to boot. Matthew Boulton was both
friend and business partner, and without him the Boulton & Watt
steam engine would not have gotten far. Watt's self-denigration does
not match the glowing terms bestowed upon him by his contemporaries. Lord Jeffrey compared his intellect to that of an elephant
trunk, which is employed with equal ease to pick up a straw or to
rend an oak. Sir Walter Scott (and many others) spoke publicly of
his kindness and benevolence. He may have felt that he could not

deal with people, but he made deep and lasting positive impressions. He also exaggerated his lack of courage. Perhaps, in a way, he exemplifies true courage by having taken a course of action *despite* being afraid. He *did* pursue the steam engine as a means to make a living, despite his fears of ruining himself and his family. And he *did* leave his home in Scotland as a teenager and travel to London to apprentice, and once there he secured the unusual arrangement of apprenticing for only one year instead of the more usual three or four. His claims of ineptitude in settling accounts may be accurate (Watt's actual words were "I would rather face a loaded cannon than settle an account"), but he also showed some business sense in the payment scheme he devised for his engines. Instead of paying for an engine outright, his mine customers, who were using the engines to pump water, paid a premium — the sum of one third of the total cost savings that resulted from exchanging an old engine for the more efficient Boulton & Watt steam-powered engine.

Watt's fearful nature may have influenced the direction of his invention. Early on, prior to his major breakthroughs with the steam engine, he was experimenting with a "Papin digester," a sort of primitive pressure cooker. To test the pressure of the steam coming out of the valve on the top, he inserted a syringe with a little piston inside it and a rod pointing out the top with some weights attached. He outfitted the contraption with a steam cock, which he could turn to control the output of the steam. In effect, he'd created a mini-steam engine, one that worked extremely well. The tiny piston inside the syringe could lift 15 pounds! This frightened Watt to the point that he dropped the line of experimentation altogether. Richard Trevithick picked the concept up decades later and built the first steam locomotive with "strong steam" in 1799.

Another way Watt's negativism manifested itself was a lifetime obsession with his health — or lack thereof. There are records of Watt complaining about poor health as early as age 19. Weariness, headaches, and other pains plagued him most of his days, and he was never particularly positive about his well-being no matter what the circumstances. In his fifties, Watt wrote, "my health is *tolerable* which is all I have now to expect." That health was a matter of major concern is evidenced by the amount of paper and ink he consumed in long descriptions of ailments, slight and significant, in his correspondence with lifelong friend Joseph Black. A letter dated April 7, 1788, is quoted in its entirety here, for full effect:

Apl 7th 1788

Dear Doctor

I am happy to find by your kind letter of the 2d that you continue to mend, and I hope you will go on in that way, and not over do yourself by too much exertion[.]

As to myself I have continued to recover, though with some retrogradations ever since Mrs. W. wrote you[.] The fever continued lurking for some time, though so as to be perceptible only by the state of my tongue the cough and a great feebleness. By proper treatment it is now quite gone, and has taken the Asthma with it, (which was a very troublesome symptom) my appetite and digestion are good and I gain strength daily. I have still however a dry cough, and a disagreeable expectoration, sometimes very salt[y], but these abate, as does a painful stitch in the fleshy part of my left breast, which has been mended by a blister. I should perhaps have sooner recovered my strength if my cons[t]itution would have permitted me to take bark, but it disagrees with me in every form and does me more harm than good. I am at present taking no medicine except a small dose of Rhubarb and tartar vitriol a[t]e twice a day, in which I have just begun[.] I have been out several times in a chaise but last week was such weather as has confined me, yet I have recovered much in that time, and hope soon to be in my usual health.

Having thus fulfilled your request, I beg leave to conclude with best wishes for yourself and Doctor Hutton

My Dear Sir yr affect[iona]t[e]

James Watt

Mrs. Watt begs your acceptance of her thanks for your kind solicitude about my health. — We hope soon to have the pleasure of seeing you here[.]

Some of the less serious diatribes on the subject of Watt's aches and pains have undertones of humor, the words of a likeable, grumpy old man. For example, in a letter to Dr. Black dated 1784 he states, "My health is as usual indifferent, and I feel I grow old and stupid. I have still however some desire for more knowledge which with the necessary attention to my family and business serves to keep me awake." A couple of years later he expressed his content-ment with life thus, "I have learnt, however, to content myself with my present negative state." His grown son once said of James Sr. that he had "never been a young man." This preoccupation with health has led at least one author to call him a hypochondriac, per-haps because Watt appeared to be quite healthy in his later years and lived to the ripe age of 83. In the end, he outlived all of his long-term friends.

Watt seemed just as fascinated with the possible cures for ill-ness as he was with illness itself. In his correspondence, long dis-cussions about the benefits of rhubarb, diet, exercise, bark, Madeira, and other remedies are juxtaposed with the results of chemistry experiments and mechanical improvements. Given to experiment as he was, he became involved with what was at the time a new method for treating sickness, that of "medicinal airs." The "pneumatic medicine" movement was based on the premise that inhaling the gases from various chemical reactions could cure people. Watt was involved in this movement in several ways. For one, he helped raise funds for a new pneumatic hospital, which served as a base for exploring this promising field of medicine as well as for treating patients. This was a serious enterprise — young science superstar Humphrey Davy was hired as an assistant to the chief doctor and medicinal air researcher, Thomas Beddoes. Watt also invented in 1794 an apparatus to isolate and administer these airs to patients. The apparatus consisted of a chamber for the chemical reaction to take place in, made of a cast-iron "fire tube," which was heated by a portable furnace. A tube extending from the chamber allowed for the addition of water, or other sub-stances, to the heated contents. A long tube (possibly made of tin[2]) was fitted to the reaction chamber, at the end of which was attached a removable bag made of oiled silk, with a wooden stopper. The "fixt" airs were collected in the bag and then breathed in by the patient through his mouth while he or she held their nose.

The apparatus was in some demand immediately, as this letter from his friend Erasmus Darwin[3] attests:

> Derby Jul 3-94
> My dear friend,
> It gave me great satisfaction both on your account and on that of the public, that you are employing your mind on the subject of medicinal airs, of which indeed Dr Beddoes had before inform'd me. You will do me a great favor by sending me an apparatus, or a description of one, and an account how easily to obtain the gasses; I have now an asthmatic patient would readily try oxygen gas mix'd with atmospheric air.

By January 1795, Watt had two sizes of the apparatus available for purchase, which he sold for the cost of the manufacture, as his interest in the matter was not about money but rather in doing his fellow mortals a good turn. He also openly encouraged imitators. In 1799, he added to the product line a breathing box, in which a patient could immerse his or her whole body as an alternative means of treatment.

Watt was also heavily involved in the testing of the apparatus. He followed very closely the experiments conducted by others into the use of medicinal airs and experimented on himself and his family. The range of chemicals tested is quite astounding: products of reactions of zinc, manganese, sulfur, iron, phosphorus, copper, burnt coals, charcoal, and slacked lime were mixed in various

---

2 In a letter to Black dated 1796, Watt told his friend that the tin used to make the tubes for the apparatus he sent are of poor workmanship. (James Watt to Joseph Black, 1 June 1796, in *Partners in Science*, eds. E. Robinson and D. McKie [Cambridge, MA: Harvard University Press, 1970], 223.)

3 Grandfather of Charles Darwin and an accomplished scientist, inventor, poet, and philosopher.

proportions in the cast-iron chamber and then heated and reheated via the portable furnace.[4] Though the experimenters did not understand the nature of the gases they were producing with these chemical reactions, and they did not have names or descriptions of them, it is likely that dropping water onto red-hot charcoal produced carbon monoxide. Dropping water onto zinc or iron filings produced hydrogen; carbon dioxide was obtained when chalk was heated, and oxygen was obtained when manganese was heated. Nitrous oxide is another possible product of the reactions Watt tried.

Results from the administration of these medicinal airs were mixed. One servant, treated by Watt himself, was apparently cured of his condition, manifested by the spitting of blood, by inhaling a mixture of steam and carbonic acid.[5] Watt described in a letter to Joseph Black how the inhalation of air from red-hot chalk with water dropped on it (likely containing carbon dioxide and carbonic acid) "causes most violent vertigo, and when very pure the barely smelling it caused a person to fall down in the sleep of forgetfulness, from which he awakened without pain or uneasiness. Dr. B. has given it in a case of incipient Phlisis,[6] which it cured."

Another young man with consumption was relieved of his chest pain through Watt's administrations of the same inhalation, though Watt noted that a weak, quick pulse and a decrease in body temperature resulted. The physician Erasmus Darwin tried oxygen on his daughter, and there were reports that oxygen also cured deafness. Watt reported in a letter to Hutton that oxygen was said to have made a 60-year-old woman menstruate, and one woman was cured of headaches after taking in 4 cubic feet (0.1 m³) of it per day. Oxygen and hydrocarbonate (carbon monoxide) gave a young

---

4 Beddoes and Davy later experimented with by-products of ammonia and discovered the effects of nitrous oxide on the nervous system. They used it to treat paralysis.

5 This was obtained from dropping water onto red-hot chalk; calcium carbonate gives off carbon dioxide at about 1,470°F (800°C). Carbonic acid is the product of a carbon dioxide and water reaction.

6 Watt likely meant phthisis, an old term for tuberculosis.

girl who normally suffered from pain with menstruation a much easier time of it. Mixtures of carbonic acid and oxygen were reported to cure ulcers and whooping cough. Others did not recover, though the reduction of pulse and full nights of sleep were considered evidence that the medicine was working.

Watt tried it on himself. When he had a pain and "coldness" in his left leg (possibly sciatica), he took four doses but no more, as it made him nauseous. It did, however, cure the "spasmodic affections"! He tried it again with success a year later to cure a cold. Watt also tried air prepared from reactions with ammonia, water, and lime, again for the bum leg. Rapid breathing and vertigo aside, it produced a "glow" in his cold leg, an effect that lasted the whole day.

While the thought of inhaling some unknown vapor may sound insane today, it was a product of Watt's time. It is easy to look through the curtain of history and proclaim Watt obsessed with health, while we sit perched on the seat of modern medicine, with its vaccines and antibiotics. But for Watt and his peers, health was a major part of everyday life; the colds and fevers that he and his friends and family wrote about every winter were often a prelude to death, also a relatively frequent topic in Watt's correspondence. In 18th-century England, an estimated 25 percent of all deaths were due to tuberculosis. By some accounts, the average life expectancy in 18th-century England was less than 40 years. Of the seven children born to Watt, only one lived to see a 50th birthday.

The medical knowledge of the time was limited. Watt's colleagues Cavenish and Priestley discovered hydrogen and oxygen, respectively. It was during Watt's time that it began to be understood that gases other than "air" existed. His friend and pen pal Joseph Black discovered carbon dioxide (though he did not identify its composition). It was a time of rapid discoveries of new elements as unknown materials were isolated for the first time. It was only then that the old idea of matter being composed of the four elements of earth, wind, water, and fire truly began to be formally challenged. The step from the discovery of gases to their use in medicine was a logical one at the time. In fact, they were trying just about anything as medicine, in desperation. Mercury was med-icine, along with various other metal concoctions. Watt also dis-cussed the benefits of living in a cow house for the treatment of consumption.

The medicinal airs inhaled by Watt and his colleagues were def-initely having an effect. In a grand twist of irony, the loss of con-sciousness and red cheeks that resulted from carbon monoxide poisoning were thought to be evidence of its benefit to the body. In a letter to James Hutton, Watt described it as a "reddner & sweet-ner of the blood." In fact, carbon monoxide is highly toxic with adverse effects on the nervous and respiratory systems. Symptoms of headaches and dizziness can result from exposures as low as 35 parts per million, while exposures of 400 parts per million for sev-eral hours can lead to death.

Dr. Lalita Bharadwaj, a toxicologist at the University of Saskatchewan, outlines the physiological effects of some of the other gases Watt and his colleagues and customers might have been inhaling:

Carbon dioxide: nausea and headache are signs of acute toxicity at exposures of 2,000 parts per million; increasing exposure will result in loss of consciousness.

Carbonic acid and ammonia: eye, skin, and respiratory irritants.

Nitrous oxide: acute exposure of over 50 parts per million can reduce dexterity, cognition, and motor and audiovisual skills. Chronic expo-sure can adversely affect haematological, immune, neurological, and reproductive systems

Metal vapors: "metal fume fever" has been observed in individuals working in welding industry exposed to metal fumes and their oxides (e.g., zinc oxide, manganese oxide). Symptoms are unspecific, but might be flu-like with fever, nausea, headache, fatigue, and joint pain.

Burnt coal: volatile organic compounds (hydrocarbons) liberated from burning coal can result in depression of the central nervous system (i.e., these are anesthetic compounds). High dose acute exposure can result in cardiac arrest.

And so Watt was quite a risk-taker after all, though, hypochondriac that he was, he would have been mortified to know it. It is a fundamental truth in the advancement of knowledge that new heights can only be reached by climbing on the scaffolding of what has come before. Watt's apparatus was a failure; the whole idea was a failure.[7] But, given the body of knowledge he had to work with, we can hardly judge him for not being able to see that.

Innovation is near-sighted.

---

7 In the 19th century, oxygen inhalation for respiratory conditions came to be recognized as a legitimate treatment with clinical benefits.

WRITE IT DOWN

ROBERT HOOKE's
FLYING MACHINE POWERED BY ARTIFICIAL MUSCLE

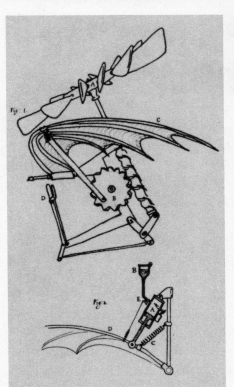

There are no existing drawings of Hooke's flying machines. This artist interpretation is based on Hooke's writings. (See endnote.)

1635–1703

London, England, in the 1660s was a crowded, vibrant, dirty place. Four hundred thousand people were crammed into its narrow, sewage-smeared streets, though the great plague of 1665–1666 reduced that number by a quarter. It was in this era that Robert Hooke, a young man in his thirties, lived and breathed science. Unlike most natural philosophers of his day, Hooke did not have independent wealth. He worked his way through the University of Oxford as a servant and afterward worked as the 17th-century equivalent of a lab technician. He was part of a small circle of men, along with Christopher Wren, John Wilkins, and Robert Boyle, who started the Royal Society in order to further science. He played an instrumental role in the Society throughout his entire life, and he spent decades working for it as curator of experiments, in charge of designing and demonstrating experiments and inventions of all sorts during meetings, often at the whim of the Society, and sometimes going without the measly pay due to him. He gave regular lectures for years, frequently to audiences of one or two: perhaps a man who'd wandered off the streets looking for a warm place to rest or a spy from his employer sent to see if he was doing his job.

Hooke had a deformed back, a natural curvature of the spine that became noticeable in his mid-teens and caused him to hunch over for the rest of his life. Coupled with his thin, wiry frame, a head that was too big for his body, thin lips, and distinctive bulging gray eyes, one can imagine that Hooke struck quite a figure as he traipsed about London on foot, avoiding fast moving carriages, horse dung, and potholes as he went. He walked a lot, with daily excursions to and from coffee houses, building sites, bookshops, and his rooms in Gresham College. Imagine we must, for there is no known portrait or other image of Robert Hooke, despite his astonishing array of contributions to science, technology, and architecture. The lack of a portrait or historic memorial to this man is a symptom of his near erasure from the common consciousness of the scientific community for over three hundred years.

Hooke was a fierce proponent of empirical science, and as such he deserves a place as a father of modern science. He supported a hypothesis/testing approach, and he fought tirelessly his entire life for his fellow men to put aside superstition and misinformation and seek truth through observation and reason:

> These infirmities of the Senses arise from double causes, either from the disproportion of the Object to the Organ, whereby an infinite number of things can never enter into them, or else from error in the Perception, that many things, which come within their reach, are not received in a right manner. The like frailties are to be found in the Memory; we often let many things slip away from us, which deserve to be retain'd; and of those which we treasure up, a great part is either frivolous or false; and if good, and substantial, either in tract of time obliterated, or at best so overwhelmed and buried under more frothy notions, that when there is need of them, they are in vain sought for . . . hence we often take the shadow of things for the substance, small appearances for good similitudes, similitudes for definitions; and even many of those, which we think to be the most solid definitions, are rather expressions of our own misguided apprehensions than of the true nature of the things themselves. . . . These being the dangers in the process of human Reason, the remedies of them all can only proceed from the real, the mechanical, the experimental Philosophy . . . there should be a scrupulous choice, and a strict examination, of the reality, constancy, and certainty of the Particulars that we admit: This is the first rise whereon truth is to begin, and here the most severe, and most impartial diligence, must be employed; the storing up of all, without any regard to evidence or use, will only tend to darkness and confusion. We must not therefore esteem the riches of our Philosophical treasure by the number only, but chiefly by the weight; the most vulgar Instances are not to be neglected, but above all, the most instructive are to be entertain'd; the footsteps of Nature are to be trac'd, not only in her ordinary course, but when she seems to be put to her shifts, to make any doublings and turnings, and to use some kind of art in endeavoring to avoid our discovery. . . . The truth is, the Science of Nature has been already too long made only a work of the Brain and the Fancy; It is now high time that it should

return to the plainness and soundness of Observations on material
and obvious things.

Hooke was arguably the first professional scientist in the world.
He believed in science and predicted its triumph, though he was
well ahead of his time. In the pure sciences his contributions
include describing capillary action, discovering the spot on Jupiter
and proving that this planet rotates, and correctly calculating the
rotational period of Mars. He explained why stars twinkle, comets
glow, and the Moon appears red near the horizon; argued that the
atmosphere gets thinner the farther away it is from the Earth; and
demonstrated how craters are formed. He surmised that gravity
emanated from the center of the celestial bodies, including Earth,
and described the formation of hailstones.

He proved that air taken in by the lungs during respiration is
essential for life; in the experiment, dogs were kept alive if bellows
blew air into the lungs directly, while a dog whose lungs were
expanded and contracted mechanically in the absence of fresh air
lost consciousness.[1] He showed that the volume of air decreases
during combustion (i.e., combustion is a chemical process) and tirelessly argued thereafter that a particular substance in air is taken up
during combustion, as well as during respiration. He, with Robert
Boyle, created gases that did not support combustion, and he figured out through experimentation that gunpowder did not need air
to combust but only ignited when saltpeter was added to the other
two components, charcoal and sulfur. He correctly surmised that
saltpeter contained the same substance in the air that causes combustion. Though he did not name it, he discovered oxygen.

Hooke wrote about the mechanical nature of memory and its
physical location in the brain; described respiration in the fetus and
the anatomy of muscle; and intuited the wave nature of light and

---

1 These experiments were highly distasteful to Hooke because of the suffering of his canine subjects,
but he recognized their importance and was forced to repeat them in order to attempt to quell the
repeated notion among other scientists that the lungs functioned to pump blood, and it was their
mechanical action that kept humans alive, not air.

sound, further exploring their properties. He studied the interaction between road surface and friction and carriage wheels, suggested the potential uses of cannabis in medicine, and established Hooke's law — that a spring stretches in proportion to the force applied to it. He experimented with water temperature and circulation, suggesting that the circulation of hot and cold currents could be utilized in a water boiler. He tested the strength of various wood species and the load-bearing capacity of ice.

In *Micrographia*, Hooke introduced the world to the microscope and the secrets it might reveal, inspiring the Dutchman Antonie van Leeuwenhoek, commonly known as "the father of microbiology." In exquisite drawings that are as stunning now as they were in their day, Hooke was the first to see a cell and coined this term, though he did not discern its function. He detailed close-ups of the eyes of a fly, the complete body of a flea, the edge of a pin, thread, crystals, the apparatus of a nettle cell's sting, insects in flight, and glass. He described how to make the microscope in enough detail that others could follow in his footsteps.

Hooke's descriptions of the behavior of light traveling through thin plates of mica inspired Isaac Newton to pursue his theories of light. The phenomenon Hooke discovered and originally described are now known as Newton's rings. He surmised that charcoal was black because it reflected no light and that different colors were due to different wavelengths, a fact that was not officially recognized until the 19th century. The subject of light would be the first nasty confrontation between Newton and Hooke, when Newton published his own works on the topic.

Hooke reasoned that crystals took their form from the internal arrangement of their particles, and he speculated how an instrument with the same structure as the nettle's sting might be used medicinally; the hypodermic injection needle was invented in 1853.[2] Hooke also suggested that the chambers of the nautilus are

---

2 In 1853, Charles Gabriel Pravaz and Alexander Wood independently invented a syringe that was fine enough to pierce the skin in order to deliver drugs. Its first application was to deliver morphine to ease pain.

filled with air for buoyancy and that amber is the preserved resin of ancient trees.

Hooke grasped that gravity played a key role in the movement of the planets before Newton, and he advocated the idea that the planets' orbits were the result of a combination of radial and tangential motion. He was the first to explain the mathematics behind the building arch. He discovered and correctly identified moss spores, stating that they were the true explanation for mosses appearing in new places, rather than spontaneous generation. The latter theory prevailed until the 19th century, when Hooke was finally proven right.

Hooke suggested that doctors use an instrument to magnify sounds produced in the body in order to make their diagnoses (i.e., the stethoscope), thought that sunspots were gases blown by solar winds, and believed that heat was due to the agitation of particles — predating the general acceptance of the kinetic theory of heat by nearly 200 years. It was Hooke who demonstrated that water expands when it freezes and that one-eighth of ice floating in water sits above the surface, revoking the commonly held belief of his day that ice sank during a thaw.

Hooke had a desire to turn his understanding of the world to the good of mankind, and hence there is a long list of inventions that resulted from the application of his experiments. He made the first working air pump in Europe, enabling important work in Boyle's laboratory on vacuums, air pressure, and combustion, including the experiments that led to Boyle's law. He devised the first glass thermometer to use the freezing point of water as zero and the first instrument to record meteorological data over time via a punched paper record.

He invented the universal joint, which is now a key component of all modern driveshafts, and the worm gear. He produced a calculating machine that could multiply up to 20 places, designed the iris diaphragm, which was reinvented later as the camera aperture, and numerous quadrants, telescopes, microscopes, timekeepers, weighing machines, and weather instruments. Some of his more

innovative instruments include a reflecting quadrant with micro-
meter adjustment, a reflecting telescope, the first Gregorian telescope, and a spring-regulated pocket watch. Yet more of Hooke's
innovations: the first machine to cut clock wheels; devices for sam-
pling sea water from the depths, measuring the depth of the ocean,
and breathing underwater; a life jacket; a submarine; diving goggles; a whale harpoon; an air-powered gun; a bubble level; an earth
auger; a cider-making machine; a machine to write in triplicate; a
kind of portable camera obscura for drawing landscapes accurately;
a means to draw huge circles for architectural applications; and a
toothed-wheel instrument that demonstrated that musical pitch is
related to frequency.

He produced vehicles — one wheeled, two wheeled, and four
wheeled. There was also survey equipment, a method for manufacturing bricks quickly, and an oil lamp with a steady, regulated oil
supply. A printing method for the type on maps is also among his
accomplishments, as well as the idea of a practical, portable book
of London maps to aid the local traveler (a forerunner to London A
to Z), the first rotary printing press, fireproof material made from
asbestos and glass fibers, and ball bearings for wheels to reduce friction. He developed a method of long distance signaling, a code for
writing the date, a horizontal mill, a contraption for walking on ice,
a means to make thunder and lightning sound effects for the stage,
various dyes for stone and wood, a waterproofing process for leather,
cleaning compounds for glass and metal surfaces, and a means to
make excellent red glass. To study motion, Hooke invented a device
with three balls suspended in a frame by wire; the very same device
has been known to grace the desk of an executive or two (and is
known as Newton's balls). It is also quite possible that the invention of the sash window is rightly attributed to Hooke.

In addition to science and engineering, Hooke had a significant
impact on London's architecture, with a list of projects that
includes the Royal College of Physicians, the Bethlem Royal
Hospital, Montagu House, the Haberdashers' almshouses, Ward's
Hospital, many of Wren's churches (St. Benet and St. Edmund the

King are most likely attributable in entirety to Hooke's designs), and collaboration on the Monument to the Great Fire.[3]

His work following the Great Fire of London in 1666 went well beyond the design of some buildings, however. As city surveyor and part of the Rebuilding Commission, Hooke, together with his close, lifelong comrade Sir Christopher Wren, had a key role in rebuilding the London phoenix that rose from the ashes. To top it all off, the dome of St. Paul's, arguably Wren's crowning achievement and in a sense the crown of London itself, was created using a method invented by Hooke.

Hooke's substantial management skill led to his involvement in a variety of municipal engineering projects, including the Fleet Canal, Thames Quay, Plymouth Royal Dockyards, and the Greenwich Observatory. He was involved in the repairs and additions to numerous other buildings, such as Westminster Abbey, as well as bridges and waterways, and he worked for a time as an agent for a builder as well. He also designed and oversaw the building of several country houses outside of London, including Escot House in Devon, Ramsbury Manor in Wiltshire, Shenfield Place in Essex, and Ragley in Warwickshire.

Robert Hooke was certainly a well-known figure in London, with friends in high places. Always though, it seems, he courted controversy. A satirical play, The Virtuoso (1676), highlighted his perceived bizarre and arrogant behavior, much to his dismay. The lead character had "broken his brains about the nature of maggots," learned to swim in the laboratory by lying on a lab table with a string tied around the waist of a frog, and attempted to turn one species of animal into another. The playwright did not think much of the new science, "As there is no lie too great for their telling, so there's none too great for their believing." Hooke wrote in his diary after seeing it, "Damned Doggs. Vindica me Deus [God avenge me]. People almost pointed."

---

3 For a complete list and images of Hooke's architecture see www.roberthooke.org.uk/arch1.htm.

His entire life, Hooke complained constantly to, and about, the Royal Society, not always unjustifiably. His pay was unsatisfactory, or withheld, and the Society's secretary conspired against him. Hooke was the sort of man that was very forthcoming with criticism. Some of his most stinging attacks were delivered via public letters or papers that were distributed among his scientific peers. He publicly accused the Royal Society secretary, Henry Oldenburg, of stealing his ideas and denounced his rivals in insulting terms.

Hooke was a braggart who was paranoid that his accomplishments would be overlooked, forgotten, and the credit given to others (with good reason, as it turned out). There seemed to be something about him, some awkwardness or chip on his shoulder, that led to him never quite getting due credit. Perhaps it was merely a result of his claiming so much. Perhaps it was jealousy, or prejudice of his appearance and background. At a Royal Society meeting in his fifties, Hooke summed up his feelings, not for the first or the last time:

> I have had the misfortune either not to be understood by some who have asserted I have done nothing, or to be misunderstood or misconstrued (for what ends I now enquire not) by others who have secretly suggested that their expectations — how unreasonable soever — were not answered. . . . And though many of the things I have first Discovered could not find acceptance yet I finde there are not wanting some who pride themselves on arrogating of them for their own — But I let that passe for the present.

One aspect of Hooke's difficult nature was his secrecy, which to others sometimes seemed to be a wrench in the gears of progress. Biographer Stephen Inwood described him thus, "his personality also had a strong element of incongruity, a degree of secretiveness, a reluctance to share insights and inventions with others, especially where profit or glory might be involved." For example, in a 1664 meeting, Hooke met with influential Londoners Sir Robert Moray and Lord Brouncker in an effort to strike a deal over a spring-regulated watch that Hooke had developed to determine longitude at sea. Such an instrument was highly sought after at the time and had great potential monetary value. Hooke showed his prospective business partners the watch but did not explain how it worked. The agreement he was offered would have provided him with ample financial reward, but Hooke would not sign it because of a stipulation that, should someone come along and improve upon his idea, the patent would be transferred to that person in his stead. Hooke broke from the deal and did not reveal his secrets. The effect of Hooke's secrecy not only stalled progress, but it also cast doubt on his later claims to the invention of the spring pocket watch.

Secrecy was not unusual in Hooke's day, as men tried to protect their claim to a discovery. For example, it was commonplace to put coded references to discoveries in a letter to record their idea prior to it being fully developed and ready to reveal to the world. Hooke's modus operandi did seem to take things a little far though; there is a pattern of him boasting huge claims of perfection and application for a nascent idea, then systematically claiming he'd already thought of everything whenever anyone else came near to furthering knowledge in the subject. On one occasion, Gottfried Leibniz demonstrated his calculating machine at the Royal Society. Hooke examined the machine carefully, with the intent of immediately one-upping Leibniz, just for the sake of it. He did so, claiming in later months to "have an instrument now making, which will perform the same effects with the German, which will not have a tenth part of the room, that shall perform all the operations with the greatest ease and certainty imaginable." In the same letter to the Royal Society, Hooke denounced the practicality of calculating machines in general. When Antonie van Leeuwenhoek announced that he'd created a microscope that could see much smaller things than Hooke's microscopes had ever done, he kept the details of how to build such a microscope to himself. Hooke immediately replicated Leeuwenhoek's result and then revealed to the world Leeuwenhoek's secret of how to build the instrument. It looks like this act was motivated first and foremost by the desire to assert his cleverness.

Hooke was involved in several very public disputes over priority and theory, once with Christiaan Huygens over the spring
pocket watch and more than once with Sir Isaac Newton, who
eclipsed him in explaining the nature of light, gravity, and planetary motion. Hooke certainly evoked frustration and rage in some
of his contemporaries. The astronomer John Flamsteed said, "He
affirms to know several secrets for the meliorating and improving of
optics, of which yet we have had no treatise. . . . Why burns this
lamp in secret?" and Johannes Hevelius complained to the Royal
Society about him: "whereas he ought to demonstrate his discoveries by observations already made by himself, he does the business
in an abundance of mere words and grandiloquent reasons . . . in
most cases it is usual with him for tasks that are to be completed to
be interrupted by other things, as though he were wholly destitute
of leisure." Huygens responded to his treatment at the hands of
Hooke during the watch dispute with a phrase describing his rival
that has a grain of truth in it: "egotistic pretension to have invented
everything."

As for Newton, he was himself an emotionally charged, overly
sensitive, and egotistical scientist; the two were bound to clash.
When Newton was elected President of the Royal Society almost
immediately upon Hooke's death, Hooke's rooms were cleared out
posthaste, and his portrait disappeared. Newton might have had
something to do with initiating the diminution of Hooke's contributions to science.

While Hooke's difficult reputation was certainly justified, it was
just as certainly exaggerated. Isaac Asimov described Hooke as a
"nasty, argumentative individual, antisocial, miserly, and quarrel-
some." This does not fully describe the man. As Stephen Inwood
demonstrated repeatedly in his biography, Hooke's life was filled to
the brim with friends who basked in his company with enough
pleasure to seek it out for decades. Hooke was a highly social creature who spent most evenings in coffee houses and taverns with his

friends. He dined with friends nearly every day, and he had long-term sexual relationships with two women,[4] one with his first house-keeper, Nell, who later married but remained a lifelong friend. The other major love interest was his niece, Grace, who came to live with him when she was 11 years old. Sexuality, most certainly secretly, entered into their relationship when Grace was 16, and she remained with her uncle in London[5] and acted as his housekeeper and companion of all kinds until her death at the age of 26.

After his death, it seems that Hooke's less pleasant attributes were emphasized, as he became best known as someone who assisted Wren and fought with Newton, though *Micrographia* and Hooke's law have always stood as testimony to his talents. It has taken over 300 years for Hooke to come into his own, and this last decade has seen a sudden resurgence in interest in the man and his work. There have been several books written about him, including: *England's Leonardo: Robert Hooke and the Seventeenth Century Scientific Revolution*; *Robert Hooke: Tercentennial Studies*; *A More Beautiful City: Robert Hooke and the Rebuilding of London after the Great Fire*; *Robert Hooke and the Rebuilding of London*; *The Curious Life of Robert Hooke: The Man Who Measured London*; *London's Leonardo: The Life and Work of Robert Hooke*; and *The Man Who Knew Too Much: The Strange and Inventive Life of Robert Hooke, 1635–1703*. A memorial to him at Westminster Abbey was dedicated in 2005.[6] There has also been an explosion of scholarly articles on topics ranging from psychology to physics concerned with

---

4 We know the details of Hooke's sex life because he recorded every occasion on which he had an ejaculation, and usually the female name associated with it, with the use of the symbol for Pisces, ✗, in the diary he kept from 1673 and 1679.

5 Grace had a 10-month sojourn at home on the Isle of Wight; it is speculated that she gave birth to an illegitimate child there, who might have been Hooke's or that of a suitor on the isle, Robert Holmes.

6 The injustice of Hooke's fate is exemplified in this monument. While the monument to Newton, featured in Dan Brown's *The Da Vinci Code*, is an elaborate sculpture, a small floor tile engraved with his name and dates are the only recognition given to Hooke — little, and late. For a photo of the monument see www.roberthooke.org.uk/memorial.htm; for comparison see Newton's monument at www.westminster-abbey.org/our-history/people/sir-isaac-newton.

his contributions, the breadth and quantity of which are yet
another tribute to his genius. The world is waking up to Hooke,
and it appears that, finally, God is avenging him.

For all Hooke's genius, and his reinstated place among the greats
in the development of science, he had plenty of failure. It was commonplace for him to promise the moon with a particular new
invention then deliver a lackluster performance. This tendency of
Hooke's to be overconfident, or overenthusiastic, followed by disappointment for the witnesses could very well have clouded his true
accomplishments. He gave promises publicly he did not keep, such
as in *Micrographia*, published in 1665, where he promised to reveal
his considerable work on human flight and his solution for measuring longitude while at sea.

Sometimes it was simply a matter of working out the bugs, such
as his ingenious depth sounder. In Hooke's day, little was known
about what lay under the sea. To rectify this, Hooke devised a contraption consisting of a float attached by a hook to a weight. The
depth sounder worked by throwing the float overboard, which sank
to the bottom. When the weight hit bottom, the float was released
and rose to the surface again. Hooke thought the time the contraption was underwater could indicate distance, based on tests of
distance covered over time, which he conducted in his laboratory.
However, he did not take into account the dynamic environment
at sea. The depth sounder did not work because the float was taken
away from the site by currents, and, when eventually found, it was
too hard to tell how long it had been underwater. Later refinements
to the instrument included revolving blades that turned at a set
rate, so the number of turns could be counted. The improved device
worked well and was reinvented in the mid-19th century.

Hooke's most novel ideas, literally hundreds of years ahead of
their time in some cases, would certainly have appeared absurd in
the 17th century, so different were they from the normally accepted
points of view. For example, his contemporaries were entrenched in
the Biblical story of the Creation and superstitious accounts of the
Earth's history. Thomas Burnet's new idea to explain the Earth's

geography (1681) was that the globe once had a uniform surface with the ocean water below it. God's flood changed all that, when He cracked the surface and let the water below burst through. In contrast, Hooke insisted that the Earth was subject to forces of change like the other planets, such as shifting rotational axes, continents and seas, and earthquakes, floods, erosion, and volcanoes: "If the Body of the Earth be accounted one of the number of the Planets, then that also is subject to such Changes and final Dissolution, and then at least it must be granted, that all the Species will be lost; and therefore, why not some at one time and some at another?"

Hooke argued tirelessly that fossils were the remains of extinct species and even hinted at evolution:

> Since we find that there are some kinds of Animals and Vegetables peculiar to certain places, and not to be found elsewhere; if such a place have been swallowed up, 'tis not improbable but that those animal Beings may have been destroyed with them . . . there may have been divers new varieties generated of the same Species, and that by the change of the Soil on which it was produced; for since we find that the alteration of the Climate, Soil and Nourishment doth often produce a very great alteration in those Bodies that suffer it . . . 'tis not to be doubted but that alterations also of this nature may cause a very great change in the shape, and other accidents of an animated Body. And this I imagine to be the reason of that great variety of Creatures that do properly belong to one Species.

Because Hooke was so far ahead of his time, his peers could not grasp the practicalities of some of his studies; for instance, Charles II laughed at his weighing of air.[7]

Hooke was a true visionary. In many cases, he had the right idea but not the technology to prove it. Sometimes a hypothesis about

---

7 To not judge Charles II too harshly, the idea of weighing air raises some eyebrows today as well, despite the advances in scientific knowledge.

a physical phenomenon was right in theory but taken to extreme. An example: he thought he could prove that the Earth moved by measuring small changes in the position of stars. While technically correct, the effect is so small that it is barely measurable, even now.

None of these explanations for Hooke's failed experiments and poor reception are to suggest that he was incapable of being flat out wrong in his observations and ideas. He was certain that life on the Moon would be discovered because he'd seen a green tinge, which led him to believe the surface of its hills were covered in a carpet of short, green grass. He also thought he had seen water on the Moon as well as one of the moons of Mars.[8] He stated that all plants are female, claimed to have made a bulletproof vest out of silk, and described shoes with springs on the bottom with which he could jump 12 feet straight up in the air. His many experiments with his own health — the taking of a multitude of substances in order to treat a multitude of minor ailments — are laughable by today's scientific standards.

Another of these flights of fancy were his flying machines. Human flight was one of Hooke's long-term interests, which he first developed as a teenager. He says that while at Winchester School[9] he impressed his headmaster most favorably with a list of more than 30 different ways to fly. He was not alone in this endeavor; many of the men in this book had an interest in the same subject, to one degree or another, with Leonardo da Vinci and Alexander Graham Bell in particular spending tremendous amounts of energy on it. Flight has been a dream of humanity since our ancestors had the psychological capacity to envy birds. Historically, it has been an emotional desire, rather than a practical one, that has driven flight; its achievement has been associated with gods and morality for thousands of years by numerous cultures, including ancient Greeks, Romans, Arabians, and Egyptians.

---

8 Mars's two moons were first discovered in 1877 by American astronomer Asaph Hall.

9 Hooke was a pupil at Westminster School from about age 13 to 18.

Not a lot survives of Hooke's human flight ideas, largely because he was secretive about the details, but we do know that the most advanced of his machine models included the following elements: wings, springs, gunpowder, and a windmill apparatus. Hooke's first attempts at building workable models took place at Oxford, post graduation, in the company of John Wilkins and others in the nascent philosophy club from which the Royal Society of London derived. John Wilkins was also a fan of flying and had written his own book in 1648 that described how man might fly to the Moon. Richard Waller, Hooke's long-standing friend, went through the papers in Hooke's rooms upon his death and subsequently published their contents as *Posthumous Works*. This note outlines Hooke's first attempts at flying machines:

> The same Year [as the air pump was made for Boyle] I contriv'd and made many trials about the Art of Flying in the Air, and moving very swift on the Land and Water, of which I shew'd several designs to Dr. Wilkins then Warden of Wadham College,[10] and at the same time made a Module [model], which, by the help of Springs and Wings, rais'd and sustain'd itself in the Air; but finding by my own trials, and afterwards by Calculation, that the Muscles of a Man's Body were not sufficient to do any thing considerable of that kind, I apply'd my Mind to contrive a way to make artificial Muscles; divers designs whereof I shew'd also at the same time to Dr. Wilkins, but was in many of my Trials frustrated of my expectations.

Waller goes on to describe several drawings in Hooke's notes showing batlike wings attached to the arms and legs of a man and a helicopter-type contraption, like a weather vane, that "by means of Horizontal Vanes plac'd a little aslope to the Wind, which being blown round, turn'd an endless Screw in the Center, which help'd to move the Wings, to be manag'd by the Person by this means

---

10 Part of the University of Oxford.

rais'd aloft." Waller judged the schemes "so imperfect" that they were not "fit for the Publick" and did not reproduce them; they are lost to us.

It is unclear whether the drawings Waller discovered were all from the same early period in Hooke's life or represented continuing work. In 1666, Hooke did experiments in front of the Royal Society that involved "winding up a spring by the force of gunpowder." Hooke called the experiment a success, with a grain and a half of powder winding up a 4-foot (1.2 m) spring to the top.

Numerous times in his diary between the years 1673 and 1679, Hooke mentioned discussions with his cronies at the local coffee houses that indicate work was ongoing, albeit sporadic. From these entries we can discern that he intended to use gunpowder to stretch the springs and that he claimed his artificial muscle could command the strength of 10 or 20 men. On October 7, 1674, he wrote that he had experimented with "artificiale strength by water, air, fire, by which flying is easy and carrying any weight," and, in contrast to earlier trials in Oxford, he was sufficiently pleased with the results that the next day he told Robert Southwell that he "could fly, not how."

Four months later, Hooke told a Royal Society Council member that he had tried both wings and kites in combination with powder. By the end of 1675, he had considered an air screw and a flying engine, which he later compared to a rowing engine, suggesting some kind of paddle mechanism. He also hinted at the use of "pulleys without wheels" and cylinders and pumps in conjunction with his flying efforts, which sounds suspiciously like a primitive piston. Hooke recorded in 1679 that he had shown a model of his design to Christopher Wren.

Throughout these years, Hooke was concurrently studying muscle anatomy, concluding in 1674 that the long, stringy fibers he saw under the microscope acted as tiny bladders, which contracted when filled with air. These investigations may have had some bearing on his designs for flight; he was certainly attempting to contrive a mechanical muscle that was based on natural muscle, as this description from the Royal Society meeting of February 3, 1670, illustrates:

> Mr. Hooke produced a contrivance of his to try, whether a mechanical muscle could be made by art, performing without labor the same office, which a natural muscle doth in animals. It was so contrived, as that by the application of heat to a body filled with air for dilatation, and by the application of cold to the same body for contraction, there might follow a muscular motion. It was objected, that it did not appear, how this agent, that was to produce heat and cold, could be applied for use, so as to cause this motion immediately, and with that speed, as it is done in animals. However Mr. Hooke was ordered to consider more fully of it, and to acquaint the society with the result of his further considerations. He suggested, that if it could be done leisurely this way, the motion might be rendered quick by springs.

Unfortunately, the snippets recorded about Hooke's flight experiments and inventions do not provide us with a clear idea of his success. However, we can still discuss the merits of some of the components of Hooke's plans, starting with the wings. While it is highly logical to approach the problem of human flight by imitating nature and turning us into a bird or bat, there are fundamental problems with this approach. While there are indeed parallels in structure between the human arm and the bird wing, there are also fundamental differences. Plastic surgeon Samuel Poore recently wrote an academic paper in the *Journal of Hand Surgery* that concludes that the human arm could be redesigned as a wing. However, said wing would be nonfunctional and serve a cosmetic purpose only.

Birds are very light for their size, in part because their bones are hollow. They also have relatively huge muscles in their breasts, which they use to flap their wings. The heavier the animal, the larger the wings need to be to support the body in the air; the larger the wings, the more muscle power required to flap them. A recent study by Katsufumi Sato broke scientific ground by attaching accelerometers, which measure thrust, to the bodies of five species of birds, including the largest flier in the world today, the albatross. The results were unsurprising in that the heavier the bird, and the longer its wings, the slower the bird can flap. Sato concluded that the upper limit of body size for winged flight is 40 kilograms. Anything heavier would not be able to keep itself in the air. The albatross weighs up to 22 kilograms, which Sato thinks fits with his calculations, as a bird closer to 40 kilograms would have difficulty flying in bad weather. Sato's conclusions suggest Hooke's attempts at outfitting a man with bat wings were a futile exercise.

Still, retired NASA engineer Paul Soderman has kindly done some rough calculations to determine just how unlikely fitting a human with wings really is. Soderman says a 140-pound (64 kg) man with 30-pound (14 kg) wings would need a 1,300-square-feet (120 m²) total wing area, corresponding to wings that are about 30 by 20 feet (9 x 6 m) each, in order to fly at a speed of 10 feet per second (3 m/s). These are pretty big wings, and it would take some pretty big pectorals to flap them up and down. It all sounds rather improbable, but it is hard to take that last step and say it is impossible. According to Sato's study, it is also impossible for the largest of the flying reptiles, such as the pterodactyl *Quetzalcoatlus*, extinct for 65 million years, to have flown. *Quetzalcoatlus* weighed in at between 200 and 550 pounds (90–250 kg) and had wingspans of up to 40 feet (12 m). While Sato claims these creatures were simply too big to fly, Soderman disagrees: "I have to side with the pterosaurs. They didn't evolve wings for nothing."

Based on the size of the wings and a person's weight alone, Hooke's plan to outfit a man with wings that flapped, otherwise known as an ornithopter, seems highly improbable, though not entirely impossible. Hooke did, at least, recognize the deficiency of human muscle as a wrench in his plan. Could his idea for a gunpowder/spring combo to provide power have worked? Senior aviation engineer Jeff Miller points out that ornithopters have thus far been failures because of "the energy requirement needed to power these machines and the physical strength needed in the construction. In short they tear themselves apart before producing enough lift." Furthermore, the use of a gunpowder explosion to compress a spring would require that the energy generated be captured within the spring. Jeff Miller says, "The main issue here is Newton's 3rd law of motion, for every action there is an equal and opposite reaction. Applied here the issue becomes how do you compress the spring? The energy of the explosion could be directed into the spring, but what about the other side of the spring? The force would travel through the spring without compressing it and directly into the wing." If Hooke's cylinders and pumps were references to a primitive engine, he was sure to fail, as the iron required to contain such an engine would have been very heavy. Gunpowder itself is a bad choice for airborne fuel as it, too, is heavy. Of course, it was an engine that eventually powered the first human flight in 1903, by Wilbur Wright, so Hooke may have been on the right track, at least in theory.

So gunpowder and springs could not have worked, and a workable engine was beyond him. However, there is enough in Hooke's own words to indicate that he was thinking in ways that were in keeping with the eventual solutions to human flight, namely the airplane and the helicopter. The breakthrough in the history of flight came with the separation of lift from propulsion; this line of thinking is attributed to George Cayley, who invented a combination of a kite or glider with paddles or a propeller — the forerunner of the airplane. Hooke was toying with kites, as well as wings, and his mention of an airscrew and horizontal vanes (a propeller) turning on an endless screw are proof positive of helicopter-type schemes.

While the airplane and the helicopter have dominated modern flight, there are other mavericks still trying to imitate birds, just like Hooke before them. Swiss pilot and aviation enthusiast Yves Rossy, otherwise known as "jet man," is one such aviator. With stiff, jetlike wings strapped to his back powered by four miniature jet engines, he crossed the English Channel at over 180 miles per hour (290 km/h) in September 2008. You can watch this birdman fly on YouTube.

Human-powered flight was made possible as early as the 1960s, with the first successful mile-long sustained flight made in 1977, in the Gossamer Condor. Simply put, this aircraft is a huge glider made of lightweight plastics with a bicycle underneath. Thus far, a human-powered helicopter, utilizing a mechanism like Hooke's airscrew, has eluded us, though it is still being tried. The most successful models have managed to get 8 inches (20 cm) off the ground. If it tickles your fancy to give it a try, there is a $20,000 prize up for grabs for the first to get 9.8 feet (3 m) off the ground and stay there for 60 seconds.

Hooke's flying machines were failures in so much as we know he did not produce a working machine that could transport a person through the air. The materials and engineering required were not yet available. However, it is possible that Hooke deserves a place in the aviation history books as one who made conceptual — if not practical — strides toward the realization of flight.

Alas, we cannot adequately judge Hooke's contribution because of his secrecy. His failure to record the full extent of his thoughts and experiments may be, historically speaking, Hooke's biggest failure of all. Hooke might properly be given credit as the first to theoretically separate lift from propulsion or to use a paddle or airscrew device in a working, flying model, but we will never know. Then again, it is just as likely that Hooke's flying machines in full-fledged form were as frivolous as other early attempts at flight, like the 13th-century flying boat with wings cranked by a handle or the 16th-century winged suit complete with a coal shovel for a tail. It was Hooke, after all, who unashamedly boasted to Sir Christopher Wren of his "flying chariot by horses."

It takes more than high-flying dreams and fancy explosions to get into the history books.

THE
GOLDEN
AGE

THOMAS EDISON's
CONCRETE
PIANO
1847–1931

The Lauter Piano Company took out a patent for a piano made of concrete and began manufacture in 1931.

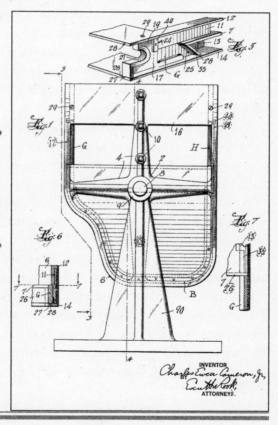

Smoke seeped through the cracks around the cellar door, adding to the acrid smell that had been lingering in the kitchen since breakfast. Young Thomas was at his experiments again. Mrs. Edison knew it was quite remarkable that her 10-year-old boy was repeating the experiments in Richard Green Parker's *Natural and Experimental Philosophy* and that the schoolmaster who'd declared Thomas "addled" had certainly got it all wrong. But surely things were a little out of hand. She and her husband really had no idea what Thomas was keeping in all those bottles and jars down in the cellar, but she had a feeling that explosions were possible. And Thomas had certainly demonstrated a fearlessness in his research that was bound to lead to trouble. Just recently she'd been embarrassed by having to send the family chore-boy home in a great deal of pain after Thomas had persuaded him to consume large amounts of Seidlitz powder.[1] He was just trying to see if the gas produced in the boy's stomach would make him float. And so, Mrs. Edison reasoned, with fire or some other catastrophe inevitable, the cellar laboratory would have to go.

Go it did, though Edison was never again far from his own lab — he seemed to build them wherever he went. On one of his first jobs, for a railway line, he secretly conducted his experiments in a freight car. Then, during the six years in his late teens and early twenties that he roamed the nation as a telegraph operator, he converted part of his office to a lab or rented a small room nearby. This was likely one of many reasons why Edison was fired, which happened often. Another reason was his getting caught using a machine of his own invention to automatically tap out the required hourly signal so he didn't have to be at his desk to do it manually. The signal was meant to signify he was awake at his desk; instead, Edison spent as much time as possible away from his desk, pursuing his own agenda — performing experiments. Later, when Edison became an independent inventor, he built his own laboratory build-

---

1 Seidlitz powder is a mixture of tartaric acid, sodium bicarbonate, and potassium sodium tartrate. It was touted in the late 19th century as a household remedy for bowel problems.

ings, choosing a spot of land near a nice place to live, Menlo Park, New Jersey, where he could walk to work, and house his employees nearby. More than one of Edison's laboratories went up in smoke during his lifetime.[2]

In addition to his predilection for setting up laboratories that would later go up in smoke and getting kicked out of jobs as he was kicked out of school, there are other traits in the young Thomas that foreshadowed his future. He was insatiably curious, both with respect to acquiring facts from book sources as well as exploring the nature of the universe with his own hands. And he was thorough. At about age 12 he set out to read the Detroit Public Library — *all* of it. He started at the bottom shelf of a rack and worked his way up to the top before moving on to the next rack. He liked to try things out for himself, as his repeat of Parker's experiments shows. That the impulse was innate, and extreme, is demonstrated by the story about how he was discovered by his father in the barn sitting on a nest of eggs to test whether he could perform the job of a hen.[3] In essence, he was a scientist — an able, mechanically minded one.[4]

From a young age, Edison also had a key ingredient for success, one that is often missing from the arsenal of creative inventors. He had a great ability with people and a way of persuading them to part with their money. He wasn't a financier. When left to his own devices during the Menlo Park days, where he invented the

---

2 Apparently, young Edison's freight car laboratory caught fire, putting an end to his activities. Much later, an explosion decimated Edison's laboratory and other buildings at his last home/work complex in West Orange, New Jersey.

3 Fiction or not, this story is characteristic.

4 Edison was a controversial figure in the scientific community of his day. There was friction, particularly with theoretical physicists, some of whom held him and his ideas in disdain. The feeling was mutual. Edison said with a truth that was sure to irk, "I can always hire mathematicians, but they can't hire me." Edison also wrote that "ordinary scientific books are in nearly every case written by men who have no capacity to explain anything," but he held some written accounts of experiments in high regard. Of Oersted's work on electromagnetism, he said, "It's awfully short, but there is literally an experiment and a fact in every line. Such is truth forever." (Edison, T.A. in *The Diary and Sundry Observations of Thomas Alva Edison*, ed. D. D. Runes [New York: Philosophical Library, 1948], 56–57.)

phonograph and the lightbulb, his bookkeeping method consisted of two hooks above his desk — one for bills, one for receipts. He was, however, a talented promoter, manager, motivator, organizer, and leader. At the age of 12 he began operating a "mobile newsstand" on the Grand Trunk Railway between Port Huron and Detroit. He had several employees and sold all manner of goods — newspapers, magazines, candy, tobacco, and fresh produce from local farmers.[5] He exploited opportunity, and his business thrived. As an adult, Edison was quick to promote his work, sometimes too quick. But if "any press is good press," then Edison knew how to get it. He frequently and readily entertained visitors in his laboratories, including parties of journalists.

When Edison was introducing the lightbulb to the world, he set up a display at Menlo Park, with strings of lights illuminating the sidewalks of the little town. Street lamps were set up in front of the railway station, and an electric chandelier shone through the windows of his parlor. First, selected journalists were invited over and treated to an incandescent-lamplight dinner. Then, it was announced that the lab would be opened to the public. Special trains ran to Menlo Park, and Edison had a public spectacle on his hands. The crowds were only permitted for a couple of weeks, however; due to the disappearance of some equipment and the breakage of a few items, the invitation was withdrawn. But the exposure had done the trick: people were captivated, and they were primed for more.

Over the next few years, while Edison worked out the numerous kinks in the plan, he continued to stage public demonstrations. Once, he had hundreds of men form a square and march down Fifth Avenue with glowing lamps on their helmets (attached by hidden wires to a steam engine/dynamo in the middle of their square). He also produced a public bulletin every 10 days or so, which outlined

---

5 Edison developed this business, and went on (as a young teen) to establish a printing press in a Grand Trunk railway car, to print his own paper, the *Grand Trunk Herald*. It was the first in the world of its kind.

progress, new customers, and gory details of gas explosions.

This ability to play the salesman did Edison a remarkable service, for, as he himself realized, "society is never prepared to receive any invention. Every new thing is resisted, and it takes years for the inventor to get people to listen to him." That is a lot of convincing. Electric lighting everywhere was a tough concept for the general public to consider feasible. Even Elihu Thomson, who was rarely wrong, publicly stated he thought the incandescent lighting system could never be profitable.

Edison's PR tendencies also had a sinister side. Edison went to great lengths to campaign for his inventions. The battle over the use of the alternate current system (invented by Nikola Tesla) versus the direct current system in mass electricity delivery is a sordid tale. One tactic Edison employed to make trouble for his enemies and their rival system was to recommend the use of the alternative current system for capital punishment by electrocution. He went so far as to assist in the experiments to demonstrate the lethality of alternative current. Of course, alternative current eventually won out; this is one occasion that Edison was wrong.

Edison's ability with people extended beyond promotion and had no small impact on the quantity of work he was able to produce during his lifetime. Edison did not work alone. He hired able men to assist him with his endeavors, and when work was heavy — like when a breakthrough was eminent or a problem required huge amounts of trial and error to solve — Edison worked these men around the clock. Edison's energy levels meant he could outwork any one of them. He was a general who fought on the front lines, sleeves rolled up, so much so that he inspired with his tireless passion and his ability to seemingly work around the clock for months on end, catching an hour or two of sleep here and there in situ. Some even said he could sleep standing up!

It was not just work, however, that kept his men so devoted. He was genuinely affable as described in early days by the *New York Tribune*:

His quaint and homely manner, his unpolished but clear language, his odd but pithy expressions charmed and attracted them. Mr. Edison is certainly not graceful or elegant. He shuffled about the platform in an ungainly way, and his stooping, swinging figure was lacking in dignity. But his eyes were wonderfully expressive, his face frank and cordial, and his frequent smile hearty and irresistible. If his sentences were not rounded, they went to the point. . . .

Edison could talk with crowds and keep his virtue and walk with kings and not lose his common touch; his contemporary Rudyard Kipling would have called him a "man."[6] He also had a boyish sense of fun and wonder and maintained a jovial atmosphere in his laboratories. He had a sense of humor — and used it. Once, when his second wife came to the lab and criticized him for spitting on the floor, offering him a spittoon instead, Edison replied that the floor was the surest spittoon because you never missed it. There was an organ at Menlo, and the group would burst into song when the mood struck. Midnight lunches were common, and whoops of excitement when a problem was solved were the norm. The sense of camaraderie was tangible; Edison genuinely loved his work, and this love was infectious. Later on in his life, Edison was asked when he would slow down; he is reported to have said, "the day before the funeral." He meant it.

Edison's work ethic was legendary and was in no small part responsible for his success. He could tackle problems requiring massive effort when others could not — or would not — because it would occupy a lifetime. With his around-the-clock team, he was able to undertake a systematic approach to problem solving: trial and error on a grand scale. For example, in his eighties, Edison acted on his wartime-inspired belief that America needed its own source of rubber. He tackled the task as he had all his other big projects, such as electric lights and their associated infrastructure. He read everything he could find of relevance. From that, he created a list of potential sus-

6 For the complete poem "If" by Rudyard Kipling, see www.kipling.org.uk/poems_if.htm.

pects, and a list of desired qualities. Then, he systematically began a research program to sift through three thousand plant species, looking for a match. This process, and the team approach Edison employed, may make perfect sense now. In fact, the approach is used broadly in R&D departments worldwide. However, it, too, can be considered an invention of Edison's — his was the original R&D lab.

It is also this process that is summed up in Edison's famous words "Genius is one percent inspiration and 99 percent perspiration." While a true reflection of Edison's modus operandi, the statement is also misleading in that it downplays his intellectual prowess. While his critics from the world of academia may have focused on his capacity for broad environmental scans and he retained a remarkable degree of modesty in his abilities his entire life, the fact is, Edison's one percent inspiration represented some heavy-duty intellectual horsepower. He was not a mathematical genius, but he was a genius nonetheless, with an uncanny ability to perceive solutions to technical problems. He may not have been the most creative of men, and many of his biggest contributions piggybacked on the ideas of others, but let's not forget that with all the book reading, data collection, and trial and error, Edison still needed to analyze and synthesize the information, apply it to new contexts, and create new solutions. He also needed to have that creative vision that all great inventors possess — the ability to see what no other has dared to see, and the belief and ability to then make that vision a reality by following it tenaciously, without regard for critics, risk of financial loss,[7] or failure.

There are plenty of examples of Edison's genius among his list of 1,093 patents. Some highlights include numerous telegraph machines and improvements to them, the first voice recording and playback system, the phonograph and subsequent improvements, the first

---

7 Edison sold his shares in General Electric and put all the money into his iron ore enterprise, which went belly up. When a colleague pointed out that Edison had not only lost $2 million, but the GE shares had since doubled in value, Edison is reported to have responded with "Well, it's all gone, but we had a hell of a good time spending it." (Clark, R.W. *Edison: The Man Who Made the Future.* [New York: G.P. Putnam's Sons, 1977], 183.)

moving-picture camera and improvements to it, and the carbon transmitter that made Graham Bell's telephone practical as well as many other telephone improvements. His most famous invention, the practical lightbulb, was accompanied by a multitude of other less-known technical devices and equipment that needed to be developed in order to produce electricity on a grand scale, deliver it steadily and safely to individual lamps — in individual rooms in individual houses and offices in the millions — and keep track of the electricity consumption. He also made significant contributions to the development of the typewriter, batteries, X-ray plates, electric cars, trains and railways, electric pens for making carbon copies, electric safety devices such as the fire alarm and warning lights, a gas mask, ore separation and processing, cement production, underwater devices for submarines and naval warfare, electroplating and other metallurgy techniques, and chemical processes, such as producing nickel hydroxide and extracting rubber. Edison also made contributions to the pure sciences, with a description of radio waves and what was known for a time as the Edison effect (the emission of electrons from a metal cathode in a vacuum),[8] though he did not follow through with either, as his plate was already full. The application of these discoveries to radio and electronics fell to others — Edison was quite happy about this and gave praise freely.

The contribution of this one man to modern life is so immense that it is difficult to appreciate. In 1931, the year of his death, the worth of his inventions to industry in the United States had mushroomed to an estimated seven billion dollars a year; that figure continues to grow as innovations are built on the foundation of his chief works. One of his ideas, the concrete piano, is, however, nowhere in sight.

How is it that a man with such well-oiled gray matter could seriously consider building a piano out of concrete? One of Edison's many talents was adeptness at applying knowledge to novel situations. So when his iron ore mining operation, run with his own ore

8 Now known as the thermionic effect.

crushing invention, ended in massive failure and left him out of pocket several million dollars, he put the machinery to use in the production of cement instead. He found himself some business partners, bought a limestone quarry, and in 1899 set up the Edison Portland Cement Company.

Here Ole Tom put his innovation and inherent business sense to work. His patented improvements to the efficiency of Portland cement production, most notably the extension of the length of the rotary kiln from a standard 60 feet (18 m) to 150 feet (46 m), helped lead to overproduction in the industry, and finding alternate uses for Portland cement became a sensible business strategy. Edison went with the flow.

Another factor influencing Edison's concrete ideas was his humanitarian streak. He thought deeply about the problems facing mankind and wrote about war, education, and poverty: "Non-violence leads to the highest ethics, which is the goal of all evolution. Until we stop harming all other living beings, we are still savages." His inventions reflected this concern for humanity. It was his humanitarianism, as well as his business sense, that led to one of his favorite ideas ever: an inexpensive concrete house that could be poured in a single operation using huge molds.

Edison reasoned that concrete, a relatively inert, strong, and highly moldable material, could be used to make prefabricated houses, furniture, tombstones, refrigerators, and even pianos. Add to that a plan for mass production. Concrete being moldable into any size or shape, Edison envisioned that affordable houses with as many as three stories — complete with bathtubs, fancy decorative ceilings, and molded window sills and other ornamental features — could be available at low cost to anyone needing a place to call home. An entire house could be built in 10 days: 6 hours to pour the concrete, 6 days for it to cure, and 4 days for the molds to be disassembled. Edison recognized the endurance of concrete as a building material, citing the Roman Baths as an example. He also recognized its low risk of fire, and other ills that beset poor folk, and the corollary lower cost of home insurance. In an interview with the *Sunday Star-Ledger* in 1906, Edison said, "If I succeed, as I feel certain I will, the cement house will be my greatest invention." The use of concrete was certainly motivated by a desire to improve the life of the common person. He said of his house, "If I succeed, it will take from the city slums everybody who is worth taking." He also declared that he would not profit from it; it was to be a phil-anthropic venture.

Edison was heavily involved in several important manufactur-ing developments that were necessary to advance the cement tech-nology of his time to enable the manufacture of houses, furniture, and pianos. Cement, before Edison, was weaker and less consistent, and it took much more time to make less of it. Though no patent was issued, Edison also developed a method of injecting air to create "foam cement," which was "lightweight," allowing for its use for household items. He figured that his foam cement could keep the weight of a house down to about 150 percent of that of wood. The concrete piano, however, was never patented. Edison, true to form, told all kinds of people about his plans for phonograph consoles, tables, chairs, and pianos made cheaply with concrete, which were soon to be commonplace in homes across America. And, true to form, the press made fun of his ideas. Cartoons appeared, and the *New York Times* joked that he'd be furnishing people with pet cats and dogs made of concrete as well. The joke was appreciated on both sides of the pond, as this quote from *Punch*, the British satiri-cal journal, illustrates:

The more extravagant party in the London County Council talk of laying liquid cement mains in Suburban London. It would be a great boon, they argue, to the ratepayer to be able to turn on the cement, just as nowadays he turns on the water for the garden hose. If unex-pected guests come for whom there is no room in the house, if a fowl house or dog kennel should be required, if the householder has ambi-tions towards a billiard room, if a porch or conservatory, or even a summerhouse should need to be built, if the roof begins to leak in a storm or (as in some cases it has done) becomes restless, if the garden

wall must be raised to keep next door from staring — in fifty different emergencies a ratepayer would find an everyday supply of cement
most useful. All he would have to do would be to send down to the
local ironmonger for the moulds, stick them up, and then leave the tap
running into them, with perhaps the youngest boy to keep an eye on
it. We would like to suggest that the cement tap ought to be coloured
red, so that it be not confused with the water tap. Cement, however
liquid, is not a good thing to water the garden with or to boil the
potatoes in.

Edison also had a deep-felt love of music, though arguably
simple taste. The role music played in his life is evident in the brief
diary he wrote in 1885, which covers a period during which he
vacationed on Nantucket Island. Piano and voice performances
were a highlight, with descriptions like "[we] lunched our souls on
a Strauss waltz." During the intense months at Menlo Park when
Edison and a crew of 70-plus employees worked on the incandescent lighting system, he hired a man to play the organ the entire
time the men were at work. The phonograph, which launched his
iconic status, was an expression of his belief that music was of inherent value in life. It was his favorite invention. Of it he said, "Music
is so helpful to the human mind that it is naturally a source of satisfaction to me that I have helped in some way to make the very
finest music available to millions." He believed, furthermore, that
to "turn America into a musical nation" was a laudable goal, and
the key to accomplishing it was "teaching music to our youth in
their homes, a different instrument to each child . . . do this and a
noble musical future is assured our coming generations." A musical
future in succeeding generations, in his view, would make a great
nation: "Think what a beautiful, marvelous America this would be
were every child taught music. With an orchestra in every family,
think what a protection to the home it would be. What sweeter or
more inspiring sight could there be than to see a family gathered
together in the lovely companionship of good music?" He even conducted experiments to "ascertain the effects of music on the human

mind," the results of which suggested to him that music has an inherent effect on the emotions. A desire to give America's children an inexpensive means to produce their own music was the basis of the concrete piano. Perhaps it was this strong sense of the good that could come from cement in the home that overrode some of his other sensibilities, like his understanding of how to give people what they want.

Edison's concrete furniture made it a little farther off the drawing board than the piano. A patent for concrete furniture submitted in 1912 was denied. The draft of this application, in Edison's hand, dated December 20, 1911, describes a steel frame with sufficient strength to withstand regular use that was "placed in moulds around which cement, preferably Portland cement is poured, to fill out the flesh so to speak of the article." The idea was that "every article of household mfr [manufacture] is capable of being made by this process and of a highly ornamental character with a cheapness unattainable by the use of wood."

That same year, Edison molded phonograph consoles from the stuff. A surviving photograph of one shows an innate piece with four clawed feet. The stand's buttresses feature acanthus leaf and grotesque elements. The phonograph compartment is octagonal, decorated with a pastiche of classical elements. The piece is topped by a water nymph, doubling its height. The nude nymph is draped in cloth that appears to have been caught in the wind and blows up over her head. Rivulets of water trickle around her feet. Edison was to show off two of his cabinets at the cement industry's 1912 trade show in New York City. But first he sent two packages with his new creation on a round-trip to Chicago and New Orleans in boxes marked "Please drop and abuse this package." No one knows what exactly happened to them, but Edison was a no-show in New York. Presumably, his pieces were not as indestructible as he thought, and he had a change of heart. He did not publicly mention concrete furniture again.

Edison's concrete houses made it furthest of all, and he bore witness to the manufacture of his dream home. A patent for the process of constructing concrete buildings, and a mold to construct said buildings, was applied for in 1908, the former patent issued in 1917 and the latter in 1915. An additional patent was issued in 1919 for an improved apparatus that allowed for the one-step building of a dwelling "in which the stairs, mantels, ornamental ceilings and other interior decorations and fixtures may all be formed in the same molding operation and be integral with the house itself." With all the technical bugs worked out, Frank Lambie and Charles Ingersoll formed the Lambie Concrete House Corporation and laid plans for a concrete subdivision in Union, New Jersey. Eleven of these houses were erected on Ingersoll Terrace, Union, in 1917. The great man himself was present for the pouring of the first one. They were put on the market for $1,200 each, in accordance with Edison's suggested retail price, approximately one third of the cost of a regular house. After a month, not a single house had sold.

Like cod liver oil, the public did not like what was good for them and ultimately rejected Edison's concrete homes. Further plans for the housing subdivision were abandoned. It seems that genius and business sense are not enough to ensure an invention's success. One must also have fashion on one's side. Edison believed that the plight of the inventor was to "try to meet the demand of a crazy civilization." In his view, failure on the public's part to share in his vision was symptomatic of their own inadequacy, rather than because of any flaw in his vision. He could be right. The foibles of crowd psychology are well recognized. Then again, it is also possible that Edison himself primed the negative response he received by spouting his ideas and plans for 11 full years before he was able to actually bring them to fruition. This provided the press with ample time to run with the story as they pleased. Perhaps things might have turned out differently if the houses on Ingersoll Terrace were the first the public had heard of Edison's ideas for concrete.

Ten of the Ingersoll concrete houses, simple cubic buildings with flat roofs, continue to house residents today (the 11th was demolished to make way for a highway exit). Several residents, interviewed for *American Heritage*, have glowing praise for the buildings, which they describe as cool in the summer, warm in the winter, with maintenance costs of "zero."

Eventually, a manufacturer did indeed take up Edison's piano vision, too. The well-respected Lauter Piano Company, which manufactured pianos from 1885 through the 1930s (and beyond) in Newark, New Jersey, produced a version of the concrete piano. The model looked for all intents and purposes like any other 5-foot baby grand. A patent for the manufacture of a piano case mold was issued in 1931. The patent describes a system whereby a mixture of materials, including sawdust and Portland cement, an inexpensive alternative to wood, are poured into molds to form the case of the piano. The inventor, Charles Ewen Cameron Jr. of the Lauter-Humano Co., describes the design as producing an "exceptionally pure, full tone."

Records of this beast are scanty, but there are several people alive who have survived an interaction with one. Kim Hunter of Orange Coast Pianos in Santa Anna, California, used to own one of them. He describes it as having standard piano keys, harp, and soundboard. It was Louis XV in style, with curved legs and music rack. Its parts were all the same thickness as a wood piano, and it was painted brown with faux finish. One thing against it: it weighed a ton, literally. Hunter says, "the piano sounded like a terrible spinet"; with no musical value it would be "better as an anchor."

Then there is the story told by Arthur Mirano, a piano technician from Florida, who worked at a piano shop in Bayonne, New Jersey, a hop, skip, and a jump from the Lauter factory in Newark. It was the late 1950s, and Mirano was 18 years old. He was restringing a piano finished in fruitwood (brown paint with flecks of black), when a fellow technician accidentally knocked a corner. A chunk of it fell off, revealing its concrete interior.

It is unknown when Lauter stopped producing this model.

It seems logical that the sound of a piano depends on the properties of its body. Whereas the wood body of a wooden baby grand resonates *with* the wooden soundboard, the concrete of a concrete baby grand absorbs the sound, deadening it. However, Professor Joe Wolfe, a physicist who specializes in music acoustics at the University of New South Wales, Australia, points out that "the properties of the case contribute little to the acoustics (apart from its geometry). For a soundboard, concrete would be very different from spruce: the former is almost isotropic, the latter very anisotropic. (Which is 'better' is of course a value judgment and lithophiles might like the instrument.)" Theoretically, then, it should be perfectly feasible to produce a concrete piano with a wooden soundboard that mimics the sound of a normal piano. And a concrete piano with a concrete soundboard that pleases at least *somebody*.

Ironically, Edison's own creations, such as the moving picture camera, phonograph, and the like, contributed to the ultimate demise of the parlor-room piano. In 1909, 374,000 new pianos were sold in America. Only one hundred thousand were sold in America in 1999, despite increased population and household income. None of them was made of concrete.

By contrast, Edison's ideas about cement houses and furniture have met a cheerier end. Concrete furniture has come into its own, with top designers working exclusively with the material to create free-form pieces of all shapes and sizes and of all palettes. Outdoor furniture is common, but indoor chairs, tables, countertops and cabinets can all be purchased online too.

As for the cement house, it is still being toyed with. In 1990, the University of Nebraska produced what they describe as "a new housing system," the Nebraska University (NU) Concrete House. Unlike Edison's one-piece affair, this is a precast concrete system consisting of exterior wall panels, floor joist panels, and an under-roof girder system. The panels are made in a factory and assembled quickly on site. As with other concrete houses kicking around, the benefits touted by its creators include low maintenance and resistance to fire, insect pests, and weather. Sound familiar?

Concrete panels poured on site are another recent construction trend. Called "tilt up," this method of construction has "proven durability attributes and low maintenance costs." According to construction industry trade publications, at least some of the benefits of Edison's proposed methods are coming into their own, albeit in altered form. Tilt up consists of flat panels that are poured in horizontal molds on site, lifted into position once cured, and held in place with strong adhesives and braces. Tilt up construction, according to *World of Concrete*, is growing in popularity as its cost-effectiveness, time-saving, disaster-resistance, and aesthetic properties become better known. It is also being toted as a green building alternative.

Could the concrete house or the concrete piano as Edison envisioned them become reality yet? It is unlikely. Even though the physical problems associated with Edison's and Lauter's piano might be overcome to produce an instrument with acceptable musical quality, the public will likely reject it, just as they have rejected the concrete sailboat, an invention with many positive features — low maintenance, strength, rot-resistance. The fact is, concrete is not, and never will be, romantic. And, apparently, most of us require romance when it comes to our boats, our houses . . . and our pianos.

It seems that the concrete piano was a rather bad idea after all, though the jury is still out if one looks at the goal of the scheme: to provide the common person with a means to create music inexpensively. Perhaps, in the early part of the 20th century, a concrete piano would have been better than no piano at all? Certainly in terms of the object's functionality, it can be said that it at least worked — it looked like a piano, and it produced sound like a piano. And if it is the thought that counts, Edison can score points with his philanthropic motivation.

But before anyone has the chance to conclude that Edison's reputation can remain unscathed, we'd best examine his other flops. There are plenty of them. His first invention, the vote recorder, was a machine to keep a vote tally in Congress, which involved calling the name of each representative in the house and manually recording their vote in the "yes" or "no" columns. Edison's invention consisted of a yes and no button installed at the seat of each representative. When the buttons were pushed to register the votes, a machine at the Speaker's desk automatically tallied the results. The device would have ensured accuracy of count and saved lots of time.

Its intended market rejected it outright. A congressional committee in Washington told the budding inventor, "If there is any invention on earth that we don't want down here, it is this. One of the greatest weapons in the hands of a minority to prevent bad legislation is filibustering on votes, and this instrument would prevent it."

There were also plans for a helicopter made from box kites and piano wire. In 1923, Edison is reported to have said, "I constructed a helicopter but I couldn't get it light enough. I used stock-ticker paper made into gun cotton and fed the paper into the cylinder of the engine and exploded it with a spark. I got good results, but I burned one of my men pretty badly and burned some of my own hair off and didn't get much further."

Edison's machine to communicate with spirits from the nether world has thus far also proved to be a dead end. The idea stemmed from a sensible enough desire to settle the question posed by spiritualists, who were quite popular in his day. Edison did not quite believe their reports of communing with the dead and thought their methods of mediums and Ouija boards were suspect. But, he was also not ready to dismiss the possibility that personalities remained intact after death. It was in keeping with his theory of the "Little People," which he wrote about at some length. Edison envisioned these invisible entities (entirely of his creation) as responsible for the functioning of our bodies, indeed of the bodies of all life-forms. He believed that the Little People have memories, vary in intelligence, and occupy all cells, though the five percent who are directors and give orders to the rest of the body inhabit Broca's folds in the brain.[9] The remainder of these tiny entities operate in highly organized units, akin to factories, perhaps even working in shifts. He believed the Little People live forever, and upon the death of their "host" leave that body, travel through the air (as well as flesh, stone, and all other matter), and enter another living being while it is an egg or seed.

---

9 Broca's area is located in the frontal lobe of the brain, roughly above and a little behind the temple. Paul Pierre Broca's work implicated this area in speech, though there is still debate about its functions, which include aspects of working memory and comprehension.

With this theory, Edison explained the relationships between all living creatures, as well as instinct — which he explained as memories carried on by the Little People — and the evolution of species (e.g., wooly mammoths became elephants when the Little People stopped producing wool). His theory also did away with the cruelty of death and predation, as a body might die, but its life force would not perish and, rather, simply be dispersed and reorganized. He was sure that life could not be created nor destroyed. Though he thought it unlikely, Edison hoped that a group of the Little People remained together as a unit after death, perpetuating one's personality. It was this collection of tiny units whose voice he thought he could magnify with his machine.

The presence of Little People was only one of Edison's faulty speculations. He also thought that freckles might be caused by iron in the skin that underwent an oxidation reaction when exposed to sunlight. It followed in his mind that freckles might be removed by a combination of the application of powerful magnets followed by "proper chemicals." He disliked education methods, understandably since he failed to conform to school as a boy and was entirely self-educated. He had a point but took the sentiment to extremes, stating "textbooks would be obsolete within a decade, replaced by moving pictures." With respect to society's ills, he felt that education was the key but was sure "you can't do anything with a grown man" and "[y]ou can't do anything or predict anything about a woman, either, because she is all instinct and emotion."

Edison recognized that "we learn a lot from our failures" and was modest enough to admit his own. When he died, most of the lights in America were extinguished or dimmed. It was a fitting symbolic tribute, not only because Edison did more than anyone to usher in the new era exemplified by that little glass bulb, but also because the lightbulb is a good representation of Edison himself: mellow, steady and bright, a conduit of energy, a manifestation of the heights of human capacity, and finite, for as far as we know, he did not send any telegrams from the other side. Like the lightbulb,

Edison is undoubtedly one of the outstanding manifestations of genius in modern civilization. There were also plenty of times when his lights failed.

Genius is not synonymous with perfection.

TRY NOT TO BE TOO WEIRD

NIKOLA TESLA's
Earthquake
Machine

1856–1943

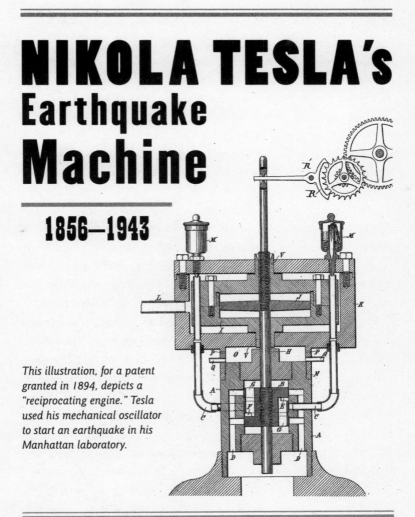

This illustration, for a patent granted in 1894, depicts a "reciprocating engine." Tesla used his mechanical oscillator to start an earthquake in his Manhattan laboratory.

"Join me in my lab at midnight, I think I can promise you some good entertainment," said Nikola Tesla to his friend Mark Twain. Twain arrived on time; he'd been to the warehouse space at 46 East Houston Street, New York, before, and he knew he'd be dazzled. His peculiar, genius of a friend was also a dramatic showman with skills rivaling Barnum. Also present that night was another friend, Chauncy Montgomery McGovern, a journalist. The lab was bright, as usual, lit by wireless lights, one of Tesla's own inventions. He found them most convenient, due to their low energy consumption, bright light, and total transportability; he carried them around and arranged them wherever light was needed. Tesla the wizard did not disappoint; purely for the entertainment of his guests he put on a pyrotechnic display, wielding balls of red flame and making his entire body glow with energy while the meter topped out at a whopping two million volts.

Twain — always the adventurer — hopped onto a square platform, about the size of a small trampoline, over to the side of the room. Tesla flipped a switch, and the platform began to vibrate silently, much to Twain's enjoyment. He whooped with pleasure, and when Tesla suggested he had had enough, responded, with the usual Twain vinegar, "You couldn't get me off this with a derrick." Tesla didn't push the point; he didn't need to. It was not long before Twain himself leaped off suddenly and asked with a panicked voice, "Where is it?" then heeded the reply by dashing straight legged to the toilet. Tesla explained that he and his assistants had themselves learned that the very pleasant physical effect of the platform was followed by a sudden, violent bowel movement.

Such displays of his inventions were not uncommon for Tesla, whose 270 or so patents in numerous countries, some of them lost, include the system of alternating current transmission of energy — used to deliver electricity to people throughout the world. The paddle-free turbine, the radio, and the induction motor — the standard technology used to power industry and household machines — are also counted among his innovations. In fact, Tesla is arguably one of the greatest inventors and scientists who ever lived. In addi-

tion to the above, he also discovered, described, and demonstrated
the fundamental concepts of some major scientific discoveries,
including cosmic rays, artificial radioactivity, X-rays, robotics, and
the electron microscope. Several of the scientists who later redis-
covered these technologies went on to win Nobel Prizes. Many of
Tesla's discoveries and inventions revolve around his understanding of matter and the existence of waves — visible light, sound,
cosmic rays, and his "very special radiation" (X-rays). At the time,
his ideas were leaps and bounds ahead of accepted explanations of
how the world works. He developed a forerunner of fluorescent
lights, drove an electric car, and publicly demonstrated a remote
control device for a model boat. He had some fairly "green" innovations too, which sound modern even today. He proposed a fertilizer system that converted, via electricity, the inert nitrogen in
the air, N_2, to the form usable by plants, N_3, and he also worked on
a system to harness energy from the Sun. He also dabbled in
weaponry, inventing death rays (or peace rays as he called them),
a form of particle gun that smacks of *Star Wars*. Extraterrestrial
radio, defensive force fields, teleportation, and time travel were
more than figments of his imagination.

The laxative platform Mark Twain rode was a mechanical oscillator. It produced vibrations by means of a relatively simple, by
Tesla's standards, engine, consisting of a spring-operated piston
encased in an iron shell and powered by compressed air or water
steam (Tesla built engines of both types). The vibrations were transferred directly to the platform but isolated from the floor via thick
cork and rubber insulating mats. Tesla was well aware of the effects
of his machine on the human body, both positive and negative, and
believed it held promise in the improvement of general health and
well-being.

The platform vibrator was one of a series of inventions that
made use of a mechanical oscillator. With this machine, Tesla
thought he could generate waves that traveled through the Earth as
sound waves travel through air. He saw many potential applications
for this technology, which he named "telegeodynamics," including

transmitting energy to anywhere on the globe and detecting the location of submarines, ore deposits, and other items of interest underground. He also thought that the patterns of waves transmitted through the globe could be read by a receiver on the other side of the world, thus forming the basis for sending messages anywhere on the Earth's surface. He imagined, then, a kind of wireless subterranean communication system.

Small, portable, and energy efficient, Tesla patented a version of the device in 1894.[1] In this patent, he mentions the powering of gears, such as that in a clock, as one application. He did not mention another possible use for the technology, that of producing earthquakes. There is no record of Tesla elaborating on the purpose that creating such earthquakes could serve; he seemed fascinated simply by the fact that he had the power to do it.

Tesla's earthquake machine was a smaller variant of the platform oscillator that "you could put in your overcoat pocket." It was about 7 inches (18 cm) long and weighed less than 2 pounds (1 kg). The small iron box contained a pneumatic piston that produced the vibrations. Tesla tested this machine in 1898 by clamping it to a steel girder in the middle of his Houston Street laboratory and starting it vibrating, and he watched the consequences. In his words:

> I was experimenting with vibrations. I had one of my machines going and I wanted to see if I could get it in tune with the vibration of the building. I put it up notch after notch. There was a peculiar cracking sound and I asked my assistants where did the sound come from. They did not know. I put the machine up a few more notches. There was a louder cracking sound. I knew I was approaching the vibration of the steel building. I pushed the machine a little higher. Suddenly all the heavy machinery in the place was flying around.

---

1 US Patent Number 514,169.

According to biographer John O'Neill, what happened was pandemonium on the streets in the surrounding area, as furniture shook, glass shattered, and panicked people flocked into the streets in response to the local earthquake. Police at the Mulberry Street station around the corner were alerted to the fact that something was afoot when their desks shook and their floor rumbled. Aware of Tesla and his antics, they naturally assumed he was the cause of any supernatural phenomena and rushed to his laboratory. Running up the stairs amid the sounds of smashing glass, they crashed through the door just in time to witness the tall, thin, and stately Tesla swing a large mallet at a small black box attached to a girder in the middle of the room. The box smashed with this first swing, and a calm Tesla politely asked the police to leave as he had more work to do. In another version of the story, the police went to Tesla's lab, but they did not witness the smashing of the oscillator. Instead, Tesla and his assistants played dumb, shrugged their shoulders, saying it must be an earthquake, and moved on with their day. Not only did Tesla later repeat this story to several newspapers, but he also went on to claim that he could use this technology to level the Empire State Building, given 5 pounds (2.3 kg) of air pressure and 10 minutes, or split the Earth in two like an apple, given a couple of weeks.

The earthquake machine was based on the principle that all materials vibrate particularly well at their own particular frequency. Applying vibrations to an object at its natural resonant frequency increases the amplitude of this vibration. This is the same effect as pushing a swing. When the force is applied to the swing at the same frequency as the swing's resonant frequency, the swing swings higher and higher with little effort. If the force is applied to the swing at a frequency that does not match, much more work is needed to increase its height. Pendulum clocks work the same way. When Tesla tried his machine out in his lab, he sent resonance frequencies out into the neighborhood that had an effect on the various buildings in his surroundings, other than his own. His own building was less sensitive, so by the time he heard the ache of protesting girders under intense stress and decided to end the experiment, plenty of neighbors were aware of the results.

Just how much damage could Tesla's earthquake machine accomplish? Engineer Isabel Deslauriers explains that the Earth does have a resonance frequency, of about 7 to 15 Hertz. However, the heavier the mass, the greater the force required to produce oscillation. Whether an oscillating device could be built with enough power to be "Earth splitting" is highly questionable. Mechanical resonance, however, can certainly be destructive. The Tacoma Narrows Bridge collapse in 1940 is an example, though this was caused by wind, rather than resonance through the ground or any attached machine. The Earth itself is not a good conductor of small waves over long distances. In fact, it has quite a dampening effect. In 2006, the Discovery Channel aired an episode of *Mythbusters* that put Nikola Tesla's earthquake machine to the test. They built several variations of the machine then experimented with metal bars to see if they could get vibrations of increasing amplitude. They had mixed results. A scale laboratory model had no effect, so a larger version of the machine was tested on a real bridge. Vibrations that were detectable 100 feet (30.5 m) away were produced, but nothing like what Tesla predicted. Jamie and Adam declared the earthquake machine a myth. You can try it for yourself if you like; a book entitled *Nikola Tesla's Earthquake Machine: With Tesla's Original Patents Plus New Blueprints to Build Your Own Working Model* is available for purchase at Amazon.com.

How is it that one of history's shining stars could expend energy on an idea of such destruction — and almost be right? The answer may have something to do with the scope of Tesla's approach to investigating the world. Tesla did not think big, he thought colossal; his imagination was stratospheres above his peers. For example, Tesla not only developed a means to provide the entire globe with free electricity by making use of energy in the atmosphere, but he single-handedly went about testing, without regard for safety, whether he was able to produce immense voltages, in the millions of volts.

Imagine the scene — he and his assistant are alone in a hangar-

like laboratory in Colorado Springs, Colorado. He has built a tower
with massive Tesla coils to magnify voltage. Imagine the assistant,
who is told to close the circuit with Tesla standing there, inviting
lightning to strike. Imagine being ordered to flip the switch then
leave the circuit closed until Tesla, who is watching from outside
the hangar, out of sight, gives the signal. Imagine standing there,
with the fear that the connection could short or the whole building blow, while massive tongues of fire arch through the air above
you. Then all goes silent and dark. During this experiment, not only
did Tesla succeed in producing lightning and the "controlled" transfer of massive amounts of energy through the air, but the power also
went out. He blew a generator at the local power station and
plunged the entire surrounding community into darkness.

Months later, Tesla returned to New York, with these experiments under his belt, in order to begin construction on a power
station that would supply wireless electricity, through antennae on
cars and buildings, to power the world at minimal cost. The lab and
Wardencliffe Tower were constructed but never completed. His
financial backer, J. Piermont Morgan, pulled out before further
experiments could get underway. Morgan had only agreed to
finance Tesla under the condition that Tesla sold the patent rights
to Morgan. Tesla had complied. Thus, he could not go to anyone
else to ask for financial backing, and the scheme was dead in the
water. It is speculated that Morgan acted intentionally to protect
his considerable investments in already existing electrical technology and its associated industry. It was not the first time that Tesla felt
he'd been "swindled." It is reported that Edison promised to pay him
$50,000 for his part in new developments to Edison's dynamos. Tesla
delivered; Edison did not. Later, Edison claimed the offer was a joke
that his recently immigrated employee did not quite understand.
George Westinghouse, founder of the Westinghouse Electric Company, wrangled Tesla out of a royalty agreement in association with
the AC patent that is estimated to have cost the inventor (and earned
Westinghouse) over $17 million turn-of-the-century dollars, equiv-
alent to a conservative estimate of about $425 million today.

Today, wireless, or "wave," energy transfer is well-known technology, with applications in radio, radar, cell phones, and Bluetooth as examples. A 21st-century company called Powercast is using radio frequencies to transmit low voltages through the air, in their words "wireless power." The range of transmission is in the realm of a dozen feet or so. So far, wireless lit-up Christmas trees and battery-free chargers for small devices that require low voltages (such as remote controls, MP3 players, cell phones, toys, hearing aids, and impact devices) are among its consumer applications. Tesla would be pleased. He said of his work, "The greatest good will come from the technical improvements tending to unification and harmony, and my wireless transmitter is preeminently such."

Tesla certainly was right about a lot of things. He was also wrong about a lot of things. His vision for wireless transmitters went well beyond modern applications. He envisioned his wireless transfer of energy would not only power households but "aerial machines . . . propelled around the earth without a stop and the sun's energy controlled to create lakes and rivers for motive purposes and transformation of arid deserts into fertile land." Chathan Cooke, principal research engineer at MIT, explains why this will never happen:

> Wave transfer of information functions well with tiny energy levels of even micro-watts. But a typical lightbulb of 100 Watts needs 100 million times greater energy transfer. This is physically possible, but there are side effects when that much energy is going in all directions. Broad area distributed, wireless electromagnetic power at levels used by a modern home will not happen. Furthermore, because of inefficiencies of traveling wave production and conversion, and our collective need for the world to be more efficient, not less, wireless power is a very local solution to specialized specific problems and will not replace wiring.

Cooke suggests one convincing reason why Tesla's idea of transmitted energy through the air is impractical: it could kill us. Life has evolved at atmospheric levels of energy in the range of 1 kilowatt per meter squared, transmitted at the wave frequencies of light. Sunlight does not kill us; however, high levels of energy at frequencies other than that of sunlight just might.

However, while Tesla's wireless energy ideas are faulty, there are plenty of other Tesla misjudgments to mull over. For example, he passionately rejected the idea that atoms were composed of subparticles, preferring the "billiard ball" model. His idea to ship mail through a tunnel under the Atlantic using water pressure turned out to be impossible, but at least he recognized his miscalculation of water dynamics. There is no such excuse when it comes to his idea to ship people around the world by means of a metal girdle surrounding the equator, built with scaffolding that is later removed. The ring would have allowed people to travel 1,000 miles per hour (1,600 km/h), the speed which the Earth spins on its axis. Somehow, the gigantic ring would have resisted the spin and stood still. Tesla's proposal for getting people on and off the ring has not survived. Neither has his description of the transporter's usefulness to the majority of the Earth's inhabitants, who do not live along the equator.

A lot about Tesla can be explained by a closer examination of how his brain worked. From a very early age, he suffered from extreme visions. In his autobiography, Tesla described their impact:

In my boyhood I suffered from a peculiar affliction due to the appearance of images, often accompanied by strong flashes of light, which marred the sight of real objects and interfered with my thoughts and action. They were pictures of things and scenes which I had really seen, never of those imagined. When a word was spoken to me the image of the object it designated would present itself vividly to my vision and sometimes I was quite unable to distinguish whether what I saw was tangible or not. This caused me great discomfort and anxiety.

Oliver Sacks has written about such altered, remarkable perceptions in patients with temporal lobe epilepsy in his book *An Anthropologist on Mars*. Tesla also complained of "flashes of light":

They were, perhaps, my strangest and [most] inexplicable experience. They usually occurred when I found myself in a dangerous or distressing situation or when I was greatly exhilarated. In some instances I have seen all the air around me filled with tongues of living flame. Their intensity, instead of diminishing, increased with time and seemingly attained a maximum when I was about twenty-five years old. . . . When I close my eyes I invariably observe first, a background of very dark and uniform blue, not unlike the sky on a clear but starless night. In a few seconds this field becomes animated with innumerable scintillating flakes of green, arranged in several layers and advancing towards me. Then there appears, to the right, a beautiful pattern of two systems of parallel and closely spaced lines, at right angles to one another, in all sorts of colors with yellow, green, and gold predominating. Immediately thereafter, the lines grow brighter and the whole is thickly sprinkled with dots of twinkling light. This picture moves slowly across the field of vision and in about ten seconds vanishes on the left, leaving behind a ground of rather unpleasant and inert grey until the second phase is reached. Every time, before falling asleep, images of persons or objects flit before my view. When I see them I know I am about to lose consciousness. If they are absent and refuse to come, it means a sleepless night.

Throughout his life, Tesla had periods of complete mental and physical breakdown. He described one of these bouts, which occurred when he was 25:

In Budapest I could hear the ticking of a watch with three rooms between me and the timepiece. A fly alighting on a table in the room would cause a dull thud in my ear. A carriage passing at a distance of a few miles fairly shook my whole body. The whistle of a locomotive twenty or thirty miles away made the bench or chair on which I sat, vibrate so strongly that the pain was unbearable. The ground under my feet trembled continuously. I had to support my bed on rubber cushions to get any rest at all. The roaring noises from near and far

often produced the effect of spoken words which would have frightened me had I not been able to resolve them into their accumulated
components. The sun rays, when periodically intercepted, would
cause blows of such force on my brain that they would stun me. I
had to summon all my will power to pass under a bridge or other
structure, as I experienced the crushing pressure on the skull. In the
dark I had the sense of a bat, and could detect the presence of an
object at a distance of twelve feet by a peculiar creepy sensation on
the forehead. My pulse varied from a few to two hundred and sixty
beats and all the tissues of my body with twitchings and tremors,
which was perhaps hardest to bear.

Tesla had few close relationships and led quite an isolated life in
many respects. He was obsessive compulsive. He was terrified of
germs, and while dining at the Waldorf Hotel restaurant night after
night, his set routine included the use of a pile of precisely 24 linen
napkins, placed to the left of his place setting, used one by one to
polish and repolish each piece of silverware. He went to extremes
to avoid shaking hands. At odds with this germ phobia was his love
of pigeons; he kept them in baskets in his hotel rooms. He found
round, smooth objects repulsive, particularly pearls, and could not
eat at a dinner party should a woman with such adornment be present. He calculated the cubic volume of each bite of food and did
everything possible in threes. He held the unshaken belief that he
would live to be 150, though he did alter that figure to 130 after
quitting whiskey in the Prohibition.

It is not uncommon in people suffering from abnormal brain
function to exhibit extreme mental capacities in particular areas.
Such is the case with Tesla. His visions led him to believe that all his
reactions, indeed all human reactions, are hardwired responses to
past experiences. He called us "meat machines" and claimed it was
this realization that led to his work on teleautomatons — known to
us now as robots. For if we are machines, Tesla reasoned, it was also
perfectly feasible that machines could be built to be like us. He had
a photographic memory and an extreme talent for mathematics. His

ability to think in three dimensions was legendary; Tesla did not tinker with his inventions. He built them in his mind, worked out the bugs, and then, once things were perfected among the little gray cells, brought his machines to life.

Creative genius is known to be associated with a higher incidence of mental illness. Tesla's brain was wired very differently from most, and while the results of that wiring have been of enormous benefit to the world, it was an anomaly, an anomaly that was accompanied by negative pathological elements. Though this is a subjective opinion, Tesla seemed to be lacking normal checks and balances when it came to making use of technology for the good of humanity, recognizing the limitations of his own mental powers, and demonstrating natural cautiousness to ensure the safety of others. Tesla tried to harness the "forces of the universe" with only a partial understanding of those forces. His words reek of delusions of grandeur:

> There manifests itself in the fully developed being — Man — a desire mysterious, inscrutable and irresistible: to imitate nature, to create, to work himself the wonders he perceives. . . . He tames the thunderbolt of Jove and annihilates time and space. He makes the great Sun itself his obedient toiling slave. Such is his power and might that the heavens reverberate and the whole earth trembles by the mere sound of his voice.
>
> What has the future in store for this strange being, born of breath, of perishable tissue, yet immortal, with his powers fearful and divine? . . .
>
> Long ago he recognized that all perceptible matter comes from a primary substance, or a tenuity beyond conception, filling all space, the Akasa or luminiferous ether, which is acted upon by the life-giving Prana or creative force, calling into existence, in never ending cycles, all things and phenomena. The primary substance, thrown into infinitesimal whirls of prodigious velocity, becomes gross matter; the force subsiding, the motion ceases and matter disappears, reverting to the primary substance.

Can Man control this grandest, most awe-inspiring of all processes in nature? . . .

If he could do this, he would have powers almost unlimited and supernatural. At his command, with but a slight effort on his part, old worlds would disappear and new ones of his planning would spring into being. He could fix, solidify and preserve the ethereal shapes of his imagining, the fleeting visions of his dreams. He could express all the creations of his mind on any scale, in forms concrete and imperishable. He could alter the size of this planet, control its seasons, guide it along any path he might choose through the depths of the Universe. He could cause planets to collide and produce his suns and stars, his heat and light. He could originate and develop life in all its infinite forms.

To create and annihilate material substance, cause it to aggregate in forms according to his desire, would be the supreme manifestation of the power of Man's mind, his most complete triumph over the physical world, his crowning achievement, which would place him beside his Creator, make him fulfill his ultimate destiny.

A sad result of Tesla's bizarreness has manifested itself in his adoption by the New Age crowd, who dress up their spirituality with pseudo-science, like the proverbial wolf in sheep's clothing. Tesla biographer Margaret Storm explained (in green "alien" ink no less) his "advanced state" as a natural consequence of his birthplace: the planet Venus. Ruth Norman, noted extraterrestrial channeler, has posited that Tesla is an alien who has come to Earth several times; one previous guise being Leonardo da Vinci. In fact, the organization Unarius, which Norman helped found, has published eight volumes of messages from the Alien Formerly Known as Tesla. These words from beyond are for sale on its website, in the "educational materials" section, with the apt title *Tesla Speaks.* In a way, it is difficult to blame Norman for adopting Tesla, as he himself spent 50 years of his life trying to communicate with Martians.

Life Technology, one company pandering to the New Age market, is selling large numbers of Tesla Shields, otherwise known as "purple plates." These companies, to varying, carefully worded degrees, attribute the invention of this technological marvel to Tesla. Purple plates, it is said, are Tesla's "personal mechanical oscillators" — his earthquake machine/platform oscillator — repackaged for a modern age. To most of us, the plates are aluminum coated in flat purple paint. To the consumers who buy them, however, they are scientific devices that "emit vast quantities of energy . . . thus the human organism can bring itself back into an equilibrium on several levels and increase its vibration frequency," bringing great health benefits. Purple plates are said to transmit positive energy to anything they touch — food, water, or furniture — cleansing any impurity. What's more, the purple paint deflects tachyons, those pesky subatomic particles that travel faster than light. These sub-particles are not only a well-known phenomenon among purple plate circles but also a definite threat to your welfare. Interesting claims indeed, considering that the physics research community is still quite tentative about their very existence. At the time of writing, purple plates line the pockets of these companies with about $100 for each 8-by-10-inch (20 x 25 cm) model sold, though a new "Hyperspace Radionics" version has been developed that sells for $179, if you wish to pay more for your purple anodized aluminum. Tesla, the unselfish humanist he was, would roll over in his grave.

While some of today's "better-informed" consumers are snatching up purple plates by the truckload, Tesla's contemporaries did not want his earthquake machine, perhaps understandably. But why they didn't jump at the chance to have wireless lights, electric cars, robots, remote control machines, or fertilizer made from air we don't know. The story of Tesla is proof that human technology is much farther ahead in the theoretical than it is in the practical. This is just as true today as it was in Tesla's day, even though some of what we could do but don't do is of obvious benefit to the globe (climate change and alternative energy comes to mind). Why we are so far behind the times at all times rests in part on the whims of big business and the powerful individuals governing what technologies receive investment — a decision that is based fundamentally on the capacity to generate cash. However, there is more to it than that. It is not just wireless lights that are resisted. We, collectively, resist anything we do not understand. We are superstitious and frightened of things that we are not born into. Without a doubt, progress is stunted by a near-universal lack of imagination. Then again, if Nikola Tesla is any example, this may not be such a bad way to be, if the alternative is fiddling with potential mass destruction and living in near insanity.

There is such a thing as too much imagination.

DON'T KILL YOUR CUSTOMERS

HENRY FORD'S
FLIPPING
FORDSON

The Fordson tractor was famous for flipping over, often with fatal results for the driver.

1863–1947

Approaching the turn of the 20th century, excitement was mounting over the horseless carriage. Numerous inventors had produced working models, including Seldon, Haynes, Duryea, Winton, Maxim, Olds, Daimler, Benz, and Marcus, before Henry Ford's tinkering in his backyard shed led to his own first car, in 1895. Ford's version had some significantly unique features. It was a gasoline-operated four-cycle design, after the Otto gasoline engine. From the beginning, Ford was thinking from first principles — not so much focusing on copying the form of steam vehicles, but rather building an entirely new vehicle from scratch. This vehicle was lighter than the other horseless carriages in existence. Its tires were not like carriage tires but like bicycle tires. Ford drove his horseless carriage around the roads of Detroit. All were in awe of the strange apparition. While Ford did not invent *the* car, he did invent *a* car, and he went on to invent numerous other cars as well. After several models — prototypes that did not meet both of his joint obsessions, perfection and mass utility — he built the Model T. The year was 1907. There were 142,000 automobiles registered in the United States. They were generally owned by the wealthy, who favored ex-coachmen as servants to drive them. Soon, however, Ford was to change the face of America with a method of mass production that would eventually have drastic consequences for the entire globe.

Ford's innovation is staggering and much broader in scope than suggested by the terms most associated with his name: "car" and "assembly line." He decided that the world needed a car that was strong and light; it would be a car that could travel on the rough roads, or where none existed, and a car that was safe and required minimal maintenance and repair. "It is my ambition," said Ford, "to have every piece of machinery, or other non-consumable product that I turn out, so strong and so well made that no one ought ever to have to buy a second one." The slogan he adopted says a great deal: "When one of my cars breaks down, I know I am to blame." Ford thought that every family needed at least one car, so he built one that was easy to operate.

Because he really *wanted* every family to have a car, Ford worked

for years to bring the price down. Completely contrary to the con-
ventional wisdom of the corporate suits, he did not maximize his
profits when setting the price point for his product. Ford's share-
holders didn't like this approach, and eventually he bought them all
out in order to have complete control over the business. He slashed
prices repeatedly, even as demand went up, all the while offering
the best quality possible. The Model T, Ford's first mass-produced
car, started at a price of $950 in 1909 and sank to $290 by 1927. For
many years, jokes about the Model T abounded, such as the fellow
who wanted to be buried with his because he'd never been in a hole
yet that his Ford didn't get him out of. One year, when his profits
were "too great," Ford mailed his customers a rebate check for $50.
Over 18 years, the Ford Motor Company sold more than 15 million
Model Ts with gross sales of $7 billion (remember this is the 1920s).
In the process, Henry Ford created an empire.

Achieving a car for the masses was a massive feat. There were
no paved roads, and the roads that existed were rough and popu-
lated with horses; horses and cars don't mix. There was no electricity in homes, no garages for repairs, no gas stations, no
dealerships, no factories. Iron and steel technology was relatively
primitive, and trains were powered by steam.[1] Ford the mechanical
genius scoured the Earth for technology that would improve his car
and make his vision a reality. His discovery of vanadium steel in a
French car at a motor race was one such breakthrough. But the real
key to his success lay in his innovative business practices. The creation of an assembly line does not begin to describe Ford's industrial
methods; he treated a vast factory system like a finely tuned watch.
Each and every square foot of the buildings, every machine, every
operation, every part, and every task were analyzed in detail and
improvements sought. Ford knew how many seconds it took for virtually every movement involved in producing a car. His plant was
designed so each operation took up minimal space. A worker's

---

1 This scenario of primitive technology and lack of infrastructure and support seems equivalent to the
status of "green" transportation today. I dare to say it: what the world needs now is a Henry Ford.

movements were minimized. All parts needed were right there, within easy grasp. The movement of parts was achieved with the use of gravity and mechanization, not bending, reaching, or lifting. The list of innovations in the Ford plant at Highland Park includes: a 5,000-horsepower gas engine, a gas-producer-regenerator-steam-boiler heat-saving system, and "over 140 special machines and several thousand special dies, tools, jigs and fixtures." There were no warehouses in Ford's holdings. To reduce the building space required, all the raw materials, parts, and finished products were in transit at all times, and the arrival of necessary materials was coordinated with production and delivery of finished goods in a relatively seamless stream of "moving inventory." Ford bought railroads, coal mines, timberlands, sawmills, glassworks, and steamships to ensure control over his dominion and avoid disruptions from others who might feel the car was a threat to their own fortune making. Quality control was second to none for the day, as were safety measures — from ventilation and good lighting to safeguards on moving parts and insistence on the use of safety goggles.

Among the most incredible, and advanced, innovations Ford instituted were his human resource policies and procedures. The background of his employees was not considered; employees were accepted regardless of past employment, criminal record, race, or culture. There was one exception: new workers were given the opportunity to indicate any special skills they might have, and any particular tasks to which they might be well suited. All this was done in order to judge a worker solely on what he could do now, not on his past performance or character. Even women, provided a husband was not supporting them, were welcome. Ford was not an angel when it came to prejudice however; he was overtly anti-Semitic, for example.

Ford believed that disabled men were also quite capable of many tasks, and he proved it. Ford calculated that of the 7,882 jobs in his factory, 3,595 jobs required no physical exertion. Of those, "670 could be filled by legless men, 2,637 by one-legged men, 2 by armless men, 715 by one-armed men, and 10 by blind men." He employed tubercular men and had a special shed constructed for those who were contagious. He had in his employ men with crippled or amputated limbs, blind men, and those who were deaf and dumb.

Ford's human resource system included minimal management — "there were no titles and no limits of authority" — and there were opportunities for every man to move up, if he so desired. He instituted systems for appeals of unfair discipline, transfers to different tasks for any worker who wanted a change, a nonprofit loan fund for employees to pay into and, if need be, borrow from, and suggestion boxes for improvements to machines and operations — the results of which were taken very seriously by the active engineering department. (When judging just how progressive Ford was in his approach, keep in mind this was nearly 90 years ago; there were no human resources laws governing his actions in these matters.)

Ford paid sick leave for his employees, made it possible for incumbents to work during recovery from their beds in order to earn wages, and — most incredibly — voluntarily more than doubled the wage of his 14,000 employees to a minimum of $5 a day. He attempted, at first, to connect the high minimum wage guarantee to healthy behavior off the job — a clean house and appearance, money management, and abstinence from addictive substances. Ford's heart was in the right place, but he later admitted publicly that he'd overstepped boundaries and retracted the conditions: "Welfare work that consists in prying into employees' private concerns is out of date."

Ford's social experiments did not stop there. In 1910, he was asked to subscribe to a new Detroit hospital that would serve its growing population. When he was asked for more money because the developers had miscalculated, Ford took over the whole thing. Ford's hospital operated on a totally different system. There were no wards; patients were grouped according to their staffing needs, akin to the machines in his factory! The service was pay as you go, each procedure's cost was made known up front, and all customers paid the same price, regardless of their status. The costs were real costs, and so the hospital paid for itself. Physicians were not allowed to have outside projects, keeping their focus entirely on their job at the Henry Ford Hospital. Several doctors, who did not confer, saw each incoming patient. Then, one physician reviewed the independent reports and, based upon these opinions, made a final recommendation for treatment. The principles were simple: minimize labor, keep everything fair, and give the best possible treatment. Although Ford and his hospital came under severe criticism from several quarters, including some physicians, the Henry Ford Hospital is today one of the leading healthcare facilities in the United States and treats more than 80,000 patients each year.[2]

Schools were another pet project of Ford's. He started several, all on a not-for-profit, self-sustainable model. There were also homes for war refugees and unemployed youth. In the early 1920s, Ford turned his attentions to history and began collecting heritage pieces of importance, which he displayed for the benefit of the public in Dearborn, Michigan. In true Ford style, the Henry Ford Museum in Dearborn is extraordinary; when Ford did something, he did it well — most of the time. Beginning with the Wayside Inn, Ford assembled a collection of important artifacts and buildings that includes Thomas Edison's Menlo Park laboratory, a 17th-century English stone cottage, a steamboat, and all manner of gadgets used by the common person.

Perhaps the penultimate of Ford's social projects was the development of small towns in rural Michigan, where he combined industry and farming in an attempt to create a utopian, self-sufficient, prosperous existence that balanced indoor and outdoor work and indoor and outdoor productivity. He also owned large tracts of experimental farmland, where he grew soybeans from which he attempted to make fiber and plastic products. Sweet potatoes, lettuce, and sugar cane were the subject of experiments in Georgia, and he financed the chemurgical laboratory of George Washington Carver.

This vast empire — assembled piece-by-piece and driven by the

2 See www.henryfordhealth.org for more information.

head of the Ford family — had a humble beginning in 1863, with Henry Ford's birth into a farming family near Dearborn, Michigan. As a child, Henry was tall and thin and disobedient. His father, William, was concerned about him: "John and William are all right but Henry worries me. He doesn't seem to settle down and I don't know what will become of him." Henry and his father were both strong-minded men with serious philosophical differences. William did not like Henry's mechanical interests; he did not like the industrial revolution, and he wanted Henry to settle down and be a farmer, just like him. Throughout his life, Henry Ford talked about the farmer in disparaging terms, describing him as a backward-thinking, inefficient sort who could make a good living if he managed his business properly, instead of working for nothing. He thought "the farmer makes too complex an affair out of his daily work" and said:

> I believe that the average farmer puts to a really useful purpose only about 5 percent of the energy that he spends. If any one ever equipped a factory in the style, say, the average farm is fitted out, the place would be cluttered with men. The worst factory in Europe is hardly as bad as the average farm barn. Power is utilized to the least possible degree. Not only is everything done by hand, but seldom is thought given to logical arrangement. A farmer doing his chores will walk up and down a rickety ladder a dozen times. He will carry water for years instead of putting in a few lengths of pipe. His whole idea, when there is extra work to do, is to hire extra men. He thinks of putting money into improvements as an expense.

Henry Ford's mechanical bent revealed itself quite early in life. As an adolescent he took to traveling around the neighborhood on horseback, fixing clocks and things for people at no charge, much to his father's chagrin. It was during this same period of Henry's life that he and his father happened upon a steam-powered road engine. Henry was fascinated and jumped off the wagon to speak to its engineer, who was only too pleased to do his best to satisfy the boy's curiosity. Again, the story goes that his father was not so pleased and tried in vain to call Henry away.

It was against his father's will that 17-year-old Henry went off to Detroit to apprentice as a machinist in a shop that made steam engines. When Henry left, he did not say goodbye. Later, William made no bones about his disapproval of Henry's dedication to the automobile. At 24, Henry did move back home, married a local girl, Clara, to whom he remained dedicated for the rest of his long life. He accepted his father's gift of 40 acres (16 ha) of timberland, which came with the condition that he give up being a machinist and settle down to log it. But the adult Henry was as disobedient as the child had been, and he built himself a first-class machine shop behind the log house. In that shop, Ford continued his quest to build the perfect car engine.

It was not long before young Ford was back in the big city again, this time working for the Detroit Edison Company. When he drove his first car, built by his own two hands, from Detroit to his father's homestead, he and his vehicle were still met with disapproval. Charles King, who accompanied him on the journey, described the reactions of father and son:

> I could see that old Mr. Ford was ashamed of a grown-up man like Henry fussing over a little thing like a quadricycle. We'd gone and humiliated him in front of his friends. Henry stood it as long as he could, then he turned to me and said, in a heartbroken way, "Come on, Charlie, let's you and me get out of here."

Ford, though he continued to believe that the average farmer was held back by his own attitude and ignorance, harbored a desire to help him. It is not surprising that very early on he formulated a plan to build and sell tractors to improve the lot of agriculturalists. Ford states that it was this desire to "lift farm drudgery off flesh and blood and lay it on steel and motors" that was the impetus for his work on the automobile. It was in 1916 that Ford's "horseless horse"[3] was finally brought to market. Called the Fordson, in honor of Ford's beloved son Edsel, the tractor caused quite a sensation. There were other tractors on the market, hundreds in fact, but the Fordson was lighter, smaller, and, most importantly, cheaper. Ford's unrealized intention was to mass-produce it in quantities that would bring the price down to $250, less than a team of horses and within the grasp of every farmer in America. He got the price to $395, but no lower.

The tractor, the Fordson Model F, was a success in many respects. In 1917, the British government begged Ford to supply them with large numbers of the vehicle. The First World War was causing massive food shortages, and there were no men at home to work the fields. Mechanization was considered a drastic but crucial solution. Ford delivered; 5,000 Fordsons were shipped to the UK at $700 each. They were driven mainly by women, who worked the British fields and saved the day. The Model F enjoyed initial success back at home too, reaching a maximum of 75 percent of the market share. More than 500,000 of them were sold on the strength of the Ford name. However, many of these customers were not happy with their purchase. In stark contrast to the Model T, the Fordson generated a lot of complaints. In 1921, there were 136 tractor manufacturers, and despite Ford's personal involvement, the Fordson was not among the best of them. Production of the Fordson Model F was discontinued in the United States in 1928 but continued in Ireland. By 1938, only five percent of all tractors bought in America were Fords. As an indicator of just how big a failure the Fordson was, relatively speaking, it was over a decade before Ford made another model in the United States, and then it was in partnership with Harry Ferguson, using the latter's patented three-point hitch technology.

The Fordson did have fans in its heyday, who appreciated the work it took off the farmer's shoulders and its savings per acre compared to

---

3 The concept of Henry Ford's tractor, the Fordson, being a horseless horse belongs to John Steele Gordon, who wrote an article entitled "Henry Ford's Horseless Horse" for *American Heritage*. The article was available online at www.americanheritage.com at the time of writing.

horses. It also has plenty of fans among today's tractor enthusiasts. However, all failure is relative. Compared to the horse, the Fordson might have been pretty good. Compared to the competition it faced in the 1920s, however, it eventually lost out. What's more, when the Fordson is judged by Ford's own standards and principles and compared to the Model T, it fails rather miserably. Ford demonstrated without a doubt his desire and capacity to give the world absolute quality, with impeccable engineering that did the job intended better than anyone, with safety of operation uncompromised. Ford saw no reason why a machine he made could not last forever because it would be made right the first time. He fully expected to be producing and selling one million tractors a year. When measured by Ford's own standards, the Fordson fails to make the grade.

There were several proximate reasons for the Fordson's failure to maintain long-term success. Its bad reputation was based on substance: it did not start easily nor go into gear easily. The first Fordsons were built with the worm gears under the driver's seat. Everything got so hot it was difficult to stay seated. Later improvements moved the worm gears back, away from the driver.

A tractor needs to do two things. First, it needs to pull equipment over land. For this, it requires pulling power and wheels that can move over rough, uneven fields. In this respect, the tractor is a horseless horse. Leaving aside difficulties in starting and driving the Model F Fordson, it can be said that it performed this duty very well. But a tractor also needs to provide the means to operate the implements it is pulling. Ford's tractor did this in the exact same way that horses do, that is, by providing the forward power to turn wheels on the implement. A horse-drawn implement, say a hay mower, has teeth on one of its wheels. As the mower's wheels rotate with traction against the ground, the teeth on the drive wheel turn a sprocket, which in turn provides the power that moves the mower blades. Implements that required more power, say grain binders, had an extra large wheel with teeth on it called a bull gear.

Farmers with horse-drawn implements who bought a Model F Fordson simply modified their implements by shortening the tongue. There were problems with the setup though. Slippage of the implement wheels as the tractor pulled it over the ground was common, and without proper rotation, the implement could not do its job. The bull gears on implements frequently got clogged and needed to be manually unclogged — a painstaking process. Of course, clogged bull gears were part of farming with horses as well as with the Fordson, but the advent of the tractor brought new challenges. In those days, operating many implements required someone to ride them. For example, a corn planter required someone to raise and lower it, and a grain binder required a rider to manually release the bundles with a foot pedal. But what did a farmer do if his children were too young to ride? A farmer could sit on his binder and steer a team of horses ahead of him with reins. Enter a problem: how to sit on the binder and drive the tractor at the same time?

On the Yesterday's Tractor forum, a fellow with the handle "Goose" recounts how he learned to drive on a 1926 Fordson. He also remembers how his father rigged up a way to drive his Fordson remotely from the seat on his grain binder with a system of ropes and pulleys: "One day one of the steering ropes broke and he started going around in circles. He pulled the rope for the clutch on the tractor and it broke, also. All he could do was jump off the binder, run and catch up with the tractor, and jump on the tractor."

It is no small wonder that an alternative means to power implements became industry standard. PTO (power take-off) bypasses the problems of ground-driven bull gears by means of a driveshaft connected directly to the tractor motor. The International Harvester Company first introduced a PTO option on a tractor in 1918. It was this innovation that eventually left the Fordson in the dust. It was not until the Ford/Ferguson duo introduced the three-point hitch, with the 9N, that a Ford tractor again held its own in the marketplace.[4]

---

4 The 9N was the first Ford tractor with the Ferguson three-point hitch; it was introduced in 1939. However, it had an inferior transmission and a few other design flaws that resulted in it being outsold by competitors. In 1947, manufacture of the 8N began. This improved model went on to become one of the best-selling tractors of all time.

PTO technology approached the problem of operating agricultural equipment from first principles — looking at the task to be accomplished and the technology available and inventing a way to do it better. The Fordson, on the other hand, was just a horseless horse. Unlike the Model T, the Fordson was not designed based on first principles to make the best machine possible for the job.

A more drastic design fault was the vehicle's tendency to flip front over back, sometimes with fatal results. A couple of agricultural rags published statistics and lists of farmers who'd been killed driving it. *Pipp's Weekly* in 1922 attributed 136 dead farmers to the flipping Fordson. Someone suggested Ford make up a decal for the vehicle for drivers to read prior to start-up, "Prepare to meet thy maker." Some of the accidents happened when a plow hit a large boulder or other obstruction that the tractor could not pull. The engine turned the sprocket, the sprocket turned the back wheels, but the tractor could not go anywhere, except up. The short wheelbase and relatively light front end of the Fordson contributed to the tendency of the front wheels to rear upward and over. Ford later added large rear fenders to the Fordson, which helped prevent this kind of accident.

A farmer named Bill, who currently chats on the Yesterday's Tractor forum, remembers an experience his bachelor uncle had with a Fordson in the United States — one of last new Model Fs with optional rear fenders:

> The Fordson was not without its little quirks. One of them being hard to start when hot. Another was having no brakes. Another was that, while idling, they had a bad habit of jumping into gear all on their own. Farmers who came in at noon for lunch and left them running would sometimes see them going across the lawn all on their own. My uncle had his Fordson sitting in the yard idling while he was working nearby. The Fordson, true to form, jumped into gear and headed straight for the silo. My uncle couldn't catch it before it plowed into the silo, climbed right up the silo and did one of its famous back flips and, in spite of being made almost completely of steel, caught fire and burned beyond

recognition. We all had a good laugh about that. My uncle went into town and bought another tractor. It was not a Ford.

Other tractor models, including modern tractors, have more stable front ends for a variety of reasons, including the placement of the implement attachment either under the tractor close to the front or up higher on the back. The Ferguson three-point hitch system, introduced broadly in Ford's 9N, also had a safety mechanism, which has been perfected in modern tractors, so that if a plow hits an obstruction it raises automatically. If Ford had applied to his tractors the adage he applied to his cars, that he is to blame every time one breaks down, he'd have had a fair amount of blood and twisted metal on his hands.

The story of the Fordson has such a different plot than that of the Model T, it seems difficult to reconcile their origins with the same innovator. Ford was a pretty simple guy, all in all, and he had some basic principles that he applied to his work and his life:

- "[E]verything can always be done better than it is being done."
- "[P]rosperity and happiness can be obtained only through honest effort."
- "[I]t is the function of business to produce for consumption . . . [which] . . . implies that the quality of the article produced will be high and that the price will be low."
- "[N]early everything that we make is much more complex than it needs to be."
- "[W]aste and greed block the delivery of true service."
- "[B]usiness exists for service."

If Ford had designed the Fordson according to his own principles, things might have turned out differently. Why did he leave his horse sense in the car factory when it came to the farmer's field? Ford says he did "not believe in starting to make it until [he had] discovered the best possible thing" and that "it is what a thing does — not what it is supposed to do — that matters," but with the first

Fordson he did not follow his own ideology. One author has suggested Ford simply didn't like farming, never did, and his subconscious resentment showed.

The complete explanation for Ford's flipping Fordson is more complicated than that. A Ford tractor was a project that he never ceased to think about; he had a prototype tractor as early as 1907. Throughout his life, he also expended a great deal of energy and money on social experiments to improve the lot of the farmer, without direct gain to himself. Perhaps the Fordson was not up to snuff because it was put on the market prematurely, in answer to Great Britain's plea for help during the First World War. However, this still does not explain why Ford's subsequent improvements did not more fully address the problems.

Perhaps, Ford was more like his father than he liked to think. It would not be the first time that father-son conflict was caused by similarity in personalities rather than differences. Ford was utilitarian to the extreme, simple, and hands-on. He rarely read as "books mussed up his mind." He applied his ideas gleaned from personal experience as a car manufacturer to everything else in his life, regardless of their relevance. Maybe he fell into the same trap he accused other farmers of doing: sticking to past methods rather than looking at the bigger picture. As he said in his autobiography, *My Life and Work*, "The farmer follows luck and his forefathers." Whatever the reason Ford disregarded his own principles, seemingly unaware; he was not the first nor will he be the last.

The yardstick you make will be the one used to measure you.

BE ON THE SIDE OF THE SUCCESSFUL

GEORGE WASHINGTON CARVER'S

Miracle Peanut CURE

. . . .

1864–1947

. . . .

George Washington Carver is shown here in his laboratory at Tuskegee Institute, Alabama.

As a Black genius living in a time and place dominated by Whites, George Washington Carver was accustomed to paradox. George Washington Carver *was* a paradox. In his heyday of fame, he was in high demand as a speaker at both White and Black institutes of higher learning. However, the same White people who had invited him to lecture then sat in rapture at his feet and given him standing ovations also forced him to sleep in separate quarters and eat alone. He's cataloged as a scientist in the annals of history, and yet he was an artist by nature and would have pursued art as a profession in all likelihood had he lived in a different time or been born a different color. He was a dreamer with an endless host of impractical ideas, and yet he was also extraordinarily pragmatic in his approach to research. He formulated questions and generated solutions that brought immediate and significant benefit to poor Southern farmers. He is widely considered a great inventor, and yet he only has three patents to his name, all of them lackluster.

"The Peanut Man's" fame is not unjustified, however. Sentiments that question whether his iconic status is merely the result of the color of his skin reek of the same prejudice that overtly held Carver back repeatedly during his lifetime. In Carver's day, the sentiment toward "Negroes" among much of White Alabama, the state in which he spent his entire professional life, can be summarized with the words "He was born for our use and our abuse, and in that place in which we have placed him he must and shall remain." All of his work was done with little financial support because Tuskegee Institute, the Black college at which he worked, received a fraction of the support of its White counterparts. Carver had neither the resources for adequate equipment nor trained hired help. His workload was immense. Under Booker T. Washington's supervision at Tuskegee, Carver's responsibilities as head of the agricultural department and the experiment station over the years included teaching four classes a day; writing and publishing several research bulletins a year; running weekend conferences for farmers and an annual six-week course for the same; taking a lead role in an annual agricultural fair on campus; attending farmers' meetings and public speaking engagements all over the South; instituting a traveling agricultural school; running two school farms, including barns, livestock, a poultry yard, a dairy, an orchard, beehives, and pastures; planning and supervising the beautification of the school grounds; looking after the school washrooms; doing analysis for the purchasing agent; attending two meetings a week on the Executive Council; and his research. On top of all of this, he also ran a Bible meeting every Sunday night; had an open door policy for all students; maintained a significant volume of correspondence with alumni, peers, and his many friends; painted; and roamed the countryside collecting samples. He was not free to follow his own professional leanings until well into his fifties, when Washington's successor allowed him the freedom to stop teaching and eventually shut down the then-redundant agricultural research facility.

As a young man, Carver was repeatedly refused entry into places of learning and wandered for about 13 years before finding a suitable place to pursue his studies, Simpson College and then Iowa State University. More than anything else, the directions to which Carver applied his genius were shaped by his experiences of prejudice, his origins in slavery, and the tense racial environment in the South. Carver was born a slave but raised as a son by his owners upon his mother's death. His life's work was almost entirely driven by a unifying philosophy and an overpowering moral obligation. Carver believed that he possessed divine gifts that he must use to help "his people." Because he had the burden of "the one great ideal of my life to be of the greatest good to the greatest number of 'my people' possible," he accepted employment at the all-Black Tuskegee Institute in Alabama rather than pursuing job offers at Iowa State or further study at the Chicago Academy of Arts. While Edison and his other contemporaries in the United States and Europe were focused on industrial processes and making the biggest impacts for the biggest number of people in the biggest way, Carver purposefully focused his attention on immediate improvements to the lives of small farmers, using local materials with no required investments of cash.

Carver was a great man with a highly inventive spirit. His innovations include over 325 products based on peanuts alone and hundreds more that used sweet potatoes, cowpeas, and other plants, which he determined, through testing, were well suited to the Southern climate and held potential for food and profit. Peanut milk; sweet potato flour; breakfast foods; coffee; paper; inks; medicines; sweeteners; formulations for paints, cosmetics, cleaners, wood stains, dyes, and wallpaper were all made from local materials with methods simple enough to be produced by local people in their homes. He created building materials and fuel from waste products, and he crossbred a prolific variety of cotton that was particularly suited to local soils and matured rapidly to avoid damage from the boll weevil. He developed over 53 products from chicken feathers, which he displayed in a fair exhibit, that he thought would prove to be "the most valuable and astonishing of any Tuskegee has attempted."

Carver's genius extended beyond the invention of physical objects to the implementation of innovative, and highly successful, modes of education. His unique three-pronged approach for his research bulletins stemmed from his lifelong employ at Tuskegee Institute's Agriculture Department. In the bulletins, Carver combined communication of his research with practical guidance for the lay farmer and recipes and other information of use to the women who ran the household.

In 1921, Carver single-handedly convinced a federal committee to put in place a tariff on peanuts to protect this growing American industry. Given 10 minutes to present his case, he began by removing "strange" objects from his pockets, which were used to illustrate his points. Extension after extension of his time was given as the committee listened, enthralled. Finally, the committee was convinced, and when he was all through, the official hearing erupted with enthusiastic applause. Not only was a tariff on peanuts written into the bill, but the committee set the value of peanuts higher than the industry ever had. Carver was a man who did things differently, and he had the charisma to pull it off.

Throughout his career, Carver repeatedly expressed a desire to have his products manufactured, but he had little success. The impetus for this desire was likely twofold. For one, it was a natural extension of his mission to improve the plight of his people. He could help the small, Southern farmer become self-sufficient with better farming techniques and nutritional improvements, but to make an economically strong South, Carver felt his people needed a broader market for their crops. He thought this would lead to a more ready acceptance of Blacks into society because of their work's economic benefit to others. Carver also likely desired personal recognition. Throughout his life, he was quick to inform his superiors (and the press) of any praise he received and in some cases requested accolades from his contacts — for example, he hinted in a letter to L.H. Pammel, his old professor at Iowa State University, that an honorary doctorate would be appreciated. It was certainly well deserved. Iowa State refused.

Despite his efforts, no Carver invention ever became part of America's compendium of common usage. Carver's hybrid cotton never became established as a major new variety. His paints, wood stains, and cleaners made from Alabama clays attracted some attention from investors and manufacturers, but the talk went nowhere. The United States Department of Agriculture once flew him to Washington to learn about his sweet potato flour, and considerable expense was put into the development of a manufacturing process for the product, but it never took off. Interest shown in his breakfast cereals petered away before the ideas became a product. His peanut milk, which he was convinced would finally let him see an idea transformed into a successful commercial venture, also died before it was born. Interest in the product was initially great, until Carver discovered that an Englishman had taken a patent out on a similar product two years previous. Carver said he'd patent his own superior process but, characteristically, never did. Carver, nor anyone else around him, pursued peanut milk further.

Carver believed in miracles. He was deeply religious, a mystic even, who believed he had visions. Upon arrival in Tuskegee,

Carver claimed he had a vision; looking out of Booker T. Washington's window at the barren clay of Alabama, he saw it transformed into rolling green hills dotted with pretty farmhouses. It is not surprising that he believed in the wonders of the peanut in miraculous proportions. The product with which he came closest to commercial success was called Penol, a medical formulation that was comprised, in part, of creosote and peanut oil, which promised to be a "Tissue Builder, Intestinal Cleanser, Germ Arrester, Nerve Food and Intestinal Antiseptic," not least of its benefits being the relief of tuberculosis. While it might sound like snake oil now, the reasoning behind Penol was sound at the time of its conception. Creosote, derived from beechwood tar, had a long history of use in medicine and was considered to be a genuine remedy for respiratory illness, cholera, diabetes, and other ails. There were several formulations on the market. Peanut oil, Carver reasoned, was a superior medium with which to spread the creosote throughout the body for maximum benefit. It was also palatable and nutritious.

Unlike all of Carver's other agrochemical products, Penol actually made it to market. With the guidance of his business manager, Ernest Thompson, and several local businessmen, the Carver Penol Company was formed and a small factory was started. The company was not particularly profitable, and Thompson made attempts to sell out cheap. The offer caught the eye of J.T. Hamlin Jr., who was selling two other medicinal remedies already, an herbal extract and a laxative. A contract was drawn up and signed that gave Hamlin the rights to produce and sell Penol in exchange for $100 a month and a modest royalty on all sales. Hamlin refitted his laboratory over the next two years to manufacture Penol and renamed his company Herb-Juice-Penol Company. Sales were disappointing, and the product never did amount to much.

There were several reasons for this. For one, the Depression had a profound influence on spending. For another, in 1937, the FDA declared Penol's nutritional claims unfounded. Then, the effectiveness of creosote in treating respiratory disorders came into question. In fact, high doses of wood creosote can cause kidney and liver damage. That said, a cough formulation containing creosote and touted as "America's first cough syrup" is still on the market. Carver did not discount the FDA's claims and remained rather silent and distanced himself from the product. One more reason for Penol's failure: it was not as palatable as Carver claimed, and he refused to improve its taste when Hamlin asked him to do so. Carver also refused to assist in the marketing of the product, or the reformulation, but would not allow Hamlin to change it either, if he wished to keep using his name.

Carver repeated a pattern throughout his career: he thought up a creative idea, advocated it as having tremendous potential, the idea garnered interest from some but not all, then it was manufactured and produced but sales never materialized. The pattern replayed itself later in the 1930s, when Carver felt he had accidentally stumbled upon the tremendous properties of peanut oil in curing polio, among other things. Fuelled by his desire to help mankind, the real benefits derived by the many ill who received his peanut oil massages, and the support of some prominent physicians, Carver tried and failed to obtain funding for testing. Again, his peanut cure was dismissed.

Penol may not have been what Carver and Hamlin claimed it was, but recent research has justified Carver's insistence of the health benefits of peanuts. The significant risks associated with peanut allergy aside, peanuts are actively being touted as a source of a number of nutrients and phytochemicals with a surprising array of positive health benefits. In addition to their high levels of vitamins E and K and magnesium and other minerals, peanuts rank equal with broccoli and tomatoes in their antioxidant properties.

Peanuts are also a source of resveratrol, which has been shown to reduce the risk of heart disease, increase energy levels and endurance, and holds promise as an effective anticancer agent and anti-inflammatory. Grapes and cocoa are better sources of resveratrol than peanuts, so a glass of red wine and some fine chocolate will have a greater effect. However, since there are other positive health benefits from peanut consumption that are not yet well understood, the best advice might be to have the red wine, chocolate, *and* the roasted peanuts too. For example, type 2 diabetes is both prevented and moderated by peanut consumption. And, although they are rich in fatty acids, the consumption of peanuts is not associated with weight gain — whole nuts being particularly low risk. In fact, it appears that peanut consumption in moderation might increase resting metabolism and thus, miraculously enough, actually help reduce weight.

Very recently, a line of research exploring the potential effects of resveratrol on neurodegenerative diseases suggest that Carver might even have been on the mark with at least some of his more miraculous claims. This compound has shown potential in the treatment of pain and tissue injury. Though yet to be tested on humans, laboratory studies have indicated that resveratrol might have a positive effect on epilepsy, Alzheimer's disease, Parkinson's disease, Huntington's disease, and nerve injury. If these tantalizing results hold true in tests on the human body, it could be that Carver was right all along, and his peanut oil, as well as his massages, was having a genuine, beneficial effect on his polio patients.

Carver's story is as much about underachievement as it is about achievement. In the annals of great inventors, his is a tragedy of Hardyan proportions, with many of his best ideas not having anywhere near the impact they could have. The reasons for this failure lie in his unwillingness to share his secrets of processes and formulas and inability to carry through his ideas to manufacturing and marketing. He lacked the dogged determination and business drive that characterized Edison, Ford, and the other greats. In Carver's own words, "I am not a finisher." He also lacked the opportunities that these other greats had, their freedom of movement, and their social power. If truth be told, Carver's inventions were most successful in garnering the public interest as fair exhibits and group demonstrations, and that must be partly attributed to his penchant for artistic display.

While ever optimistic, Carver must have been frustrated by this failure. There is not a single concrete example of how society has benefited greatly from his inventions. Many of his creations were never more than ideas, and many of his products were not particularly innovative. And yet, like the heroes Hardy penned, his life's work serves to remind us of society's topsy-turvy priorities; for while Carver's inventions did not have any significant impact on the commercial world, his work had immeasurable impact on the individuals around him.

Penol was undoubtedly a failure as an invention; it did not do what it was intended, and the public did not like it. If success is measured by commercial profit or mass utility, the rest of Carver's inventions are also failures. However, if we take into account the broader context of Carver's work and his grounding purpose — to lift up the small Southern farmer and assist his people, Carver is a resounding success. Through his outreach on all levels, he personally touched the lives of tens of thousands and has affected countless more ever since. One of those whom he influenced, Howard Kester,[1] wrote, "Marvelous are the miracles you have performed in the laboratory but more marvelous still are the miracles you have wrought in the mind and heart of hundreds of men and women." When judging George Washington Carver's success as an inventor, and a man, we face a clear choice: to measure success by dollar value or by, the less readily quantified, impact on human lives.

Success is defined by the successful.

---

1 Howard Kester was a preacher and activist; he organized the Southern Tenant Farmers Union in 1934.

BE SURE THE MARKET IS READY

ALEXANDER GRAHAM
BELL'S
Six-Nippled Sheep

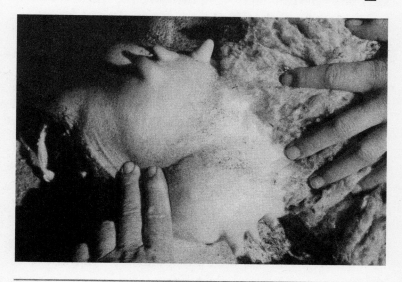

George Davis, agricultural researcher in New Zealand, is right now engaged in a multi-nipple breeding program, which includes the six-nippled ewe on display here.

1847–1922

"It is astonishing how ignorant we all are about common things. Just test the matter on yourself. Sheep are quite common; and we are all more or less familiar with their appearance, and should therefore be able to answer some questions about them. Well then . . . how many nipples has she, and where are they located?" So asked Alexander Graham Bell of *Science* readers. Bell himself was made aware of his own ignorance on the matter in 1890, when he examined the flock of sheep that came with the property he purchased on Cape Breton Island, Nova Scotia. The examination taught him that sheep have two nipples and that they are positioned under the hips, like those of a cow. The occasion also marked the onset of Bell's longest standing experiment, which entailed a selective breeding program for multi-nippled sheep. But the origins of this work lie several years previous, in 1886, when Bell, his wife Mabel, and their two young daughters bought a summerhouse in the village of Baddeck, Cape Breton. A pet lamb was purchased to go with it. After a pleasant summer, the Bells returned to Washington, DC, for the winter. When they returned to Baddeck the following spring, they now owned two sheep — last summer's lamb had turned mother.

Why only one lamb, wondered Mrs. Bell and her husband. Specifically, Mrs. Bell reasoned, "A progeny of nineteen was no uncommon event in a pig's family. Even dogs generally had as many as six at a birth, while twin lambs were rare, and quadruples unheard of. . . . Neither the pig nor the dog mother had difficulty in rearing a large proportion of their children to fine maturity. Why could not the sheep do so too? The problem fascinated Mr. Bell."

That the inventor of what is now known as the single most valuable patent in history, the telephone, could spend 32 years attempting to breed sheep that had litters of progeny might come as a surprise, but not to those who knew the man. Bell had a Scottish countenance and disposition. He had massive black eyes, a massive shock of hair, massive features, and a massive personality to go with them. He was a dramatist with an insatiable curiosity about everything around him and a genius. He was well married

— undoubtedly for love — but his wife also brought him independent wealth. So when the Bells bought their Cape Breton headland in 1889, along with surrounding farmland, and built the mansion Beinn Bhreagh, he was able to let his curiosity soar and made serious work of tinkering for the rest of his life: "There seems to be always something going where I am. Nothing, perhaps, that would interest other people, but it keeps me busy and interested all the time."

Bell's tinkering had some impressive results. Other than the telephone, he had plenty of good ideas that, over time and with subsequent advances in knowledge and technology, proved to be extraordinarily useful. He developed an artificial respiratory device, a forerunner to the iron lung that was developed in the 1920s; he foresaw the demise of fossil fuels as a reliable energy source and the importance of solar power, and he perceived "the great hope of the future . . . to lie in alcohol, a beautifully clean and efficient fuel." He experimented heavily with the use of a beam of light to transmit the spoken word and considered his "photophone" to be his greatest discovery, outranking the telephone in importance. He developed a means to use sound waves to locate items underwater and began to think about a means of safe artificial birth control. He published a paper on the use of radium in sealed glass tubes for the treatment of deep-seated cancer, an idea that has recently been revisited in an altered form. Bell believed that the tetrahedral shape could be instrumental in the design of safe, stable aircraft as well as buildings — technology that is now used in building bridges and roofs under the colloquial name "space frame." He also made great strides toward a flying machine, using tetrahedral structures in the building of giant kites.

Bell loved nature. He and his family spent a great deal of their time at Beinn Bhreagh, often returning to Washington for only a month or two each year. The wilderness suited Bell, who was described by his daughter Daisy as quite an "elemental person, a creature of the water and the woods and the night." He preferred to relax in *puris naturalibus*, and consequently stories abound of Bell being caught without his clothes on, asleep on his houseboat/study, which was moored at the edge of the lake. On occasion, he was lured into the water by the moonlight for a skinny dip (though there was nothing skinny about him) and would subsequently find himself unable to locate his knickerbockers. Wading through snow with naught but a bathing suit and winter boots for warmth seems to have filled Bell with glee, and once, as an experiment, he put himself out in the wild armed only with a penknife and his wits with which to find clothes, food, and shelter. After trying out some damp sphagnum moss as a loincloth, Bell concluded that "the scientific man would get on better in some other climate than Canada."

Bell loved animals and kept a few bobcats, bald eagles, and a bear. His interest in sheep, while genuinely scientific, was also more than that. The flock he acquired in 1890 was kept in a carefully planned "sheep village" called Sheepville, complete with main and cross street signs. He liked to have his sheep close by and kept a box of oats at the gate to the sheep pasture, close to the house. He and his grandchildren were in the habit of feeding the sheep in the afternoon, and "he was delighted in having his flock tame enough so that they would eat from his hand." In 1914, when Bell, frustrated with the lack of progress in his selective breeding, sold the flock, he "felt rather lost," and Mrs. Bell secretly bought some back. Mrs. Bell's flock, as they were known from then on, continued to be a source of enjoyment for the inventor and his family for the rest of his life.

Bell's sheep breeding might be considered a gentleman's hobby, but it was also a serious endeavor, and he kept highly detailed notes of all breeding in his *Beinn Bhreagh Recorder*, self-described as "a type-written periodical limited to five copies," containing all manner of notes and records of the conversations and experiments that took place at his Cape Breton home. Bell sent a copy of this scientific journal/personal newsletter to the Smithsonian Institution for posterity. His reports included the carefully kept life history data of every sheep in his flocks, including sex, color, number of nipples, single or twin parentage, and progeny. Births, weights, and matings were also tabulated by hand exhaustively and analyzed in depth.

In the first year of lambing, in 1891, one half of the flock's off-spring were twins. Several ewes in the flock had one or two extra, non-functioning, nipples. Twenty-four percent of ewes with the normal two nipples had twins, while among those ewes with rudi-mentary supernumerary nipples, 43 percent had twins. Bell excit-edly wrote his wife with the results, and he proceeded with a selective breeding program to attempt to increase the number of functional nipples to four. He hoped that ewes with extra mam-mary glands would also have more offspring. He scoured the coun-tryside for multi-nippled sheep. He put up notices offering several times the market value for sheep with extra nipples — the more nipples, the more cash. And he made progress.

In 1904, Bell reported to the National Academy of Sciences that he had successfully bred a flock with four functional mammae yielding milk and had noted lambs with five and six nipples in his flock, including one ewe with four nipples on one side of her body and two on the other, "foreshadowing the possibility of an eight-nippled variety." He reported that twin production was low and not more prominent among multi-nippled sheep, but that his sample sizes were very small and his multi-nippled ewes were young. Young ewes are less likely to twin than older ewes. There was also sugges-tion that the twins, though smaller at birth, speedily came up to the average of the flock and weighed as much as their single lamb counterparts by autumn. To Bell, this signified that multi-nippled sheep could very well be able to rear twins more effectively than "regular" sheep, and he was encouraged to continue with his breed-ing experiments, this time focusing on producing not only a multi-nippled flock but a multi-*offspring* flock. To do this, Bell recognized that he needed to increase the size of his flock and introduce new blood. Having exhausted the supply of naturally occurring super-numerary-nippled sheep on Cape Breton, Bell formulated plans to breed normal two-nippled ewes to his multi-nippled rams. An inventor's solution: if you can't buy it, build it yourself.

In 1912, Bell reported that though rare in nature, six-nippled sheep do occur occasionally. By selective breeding and outbreeding, Bell had increased his stock of ewes with six functional nipples. Over 50 percent of the previous season's lambs had been six-nippled. He related his plan to cut down his flock to include only ewes with six functional nipples.

His experiments to increase twinning proved more complicated, as Bell grappled with learning the factors associated with bearing multiple offspring: age of ewe, season, size and condition of ewe, nutrition during mating season, and nutrition during pregnancy. Bell had learned that his original idea — that a sheep with more nipples will have more offspring — was wrong. However, he showed that nipple number is a flexible, heritable trait. And he was optimistic that, by selecting twinning ewes in fine physique that gave birth to small lambs that subsequently grew quickly, he would be able to produce a breed of sheep that had two offspring every season and lots of milk to feed them. A far cry from his original vision of a sheep litter, but an accomplishment nonetheless.

Alexander Graham Bell died in 1922. His wife published an update on the sheep breeding experiments posthumously, with the tragic subtitle "Mr. Bell's Last Contribution to Science." She herself died of a broken heart before the article went to print. As the article outlines, with Bell's six-nippled flock sold in 1914, the remnants of the flock purchased back by Mrs. Bell were the primary focus of twin-bearing experiments. It seems that the six-nippled flock had not met Bell's expectations, for the six-nippled trait was not so established that all lambs born of male and female six-nippled parents had six nipples themselves. Mrs. Bell was satisfied with four functional nipples, however, and she had refocused their efforts on ensuring that all supernumerary nipples were as large and well developed as the primary pair. Bell admitted that he had, in the past, been "too anxious about the mere number of nipples" and stated that this "doubtless had been the reason for our failure to establish the six-nippled tendency as a hereditary trait." By 1922, this small flock could have been described as twin-bearing as well as possessing four well-developed, functional nipples. Of 31 lambs born that year, one was a single, 24 were twins, and six were triplets. What is more, by the fall the twins and triplets were fully equal in weight to the single lambs. Bell recognized that he could not say with certainty whether the multiple-offspring trait was hereditary or not, but he concluded that the experiments to find out were certainly warranted.

Bell's notes were examined in 1924 by W.E. Castle, a Harvard geneticist, who determined that the multi-nippled trait was strongly established in the flock, but the presence of increased milk production and twinning tendency were not so clear. The University of New Hampshire Experimental Station took on the flock and continued with the experiments. In 1942, *Time* magazine published a short update on the status of the flock as it changed hands; the retiring professor in charge passed them on to the United States Bureau of Animal Industry. The United States Department of Agriculture subsequently concluded that "the multi-nipple character has no practical value in sheep production" and ended the experiments. Six-nippled sheep, even the *idea* of six-nippled sheep, fell into obscurity.

Bell's sheep breeding work was a failure on multiple levels. His original intent of creating a breed of sheep that gave birth to large litters never came to fruition. His consequent attempts to breed sheep that produced twins that grew quickly to the size of singles were inconclusive. Some of this failure can be blamed on gaps in knowledge rather than faulty logic. The fields of genetics and population biology were not well developed in Bell's time. He proved that breeding for multiple functional nipples was perfectly feasible. He failed, however, to understand the link — or lack thereof — between mammary gland number and litter size. Both traits are heritable and associated but are genetically isolated. Large numbers of nipples does not a large litter make.

The evolution of lactation predates the evolution of mammals and appears to have its origins in apocrine glands associated with hair follicles. Cynodonts, mammalian ancestors in the Triassic Period (over 200 million years ago), almost certainly secreted nutrient-rich milk to feed their hatchlings. This is inferred from the small body size of the hatchlings and teeth formation, among other things. Monotremes, the branch of mammals whose list of living representatives is limited to the echidna and the platypus, do not have nipples but rather clusters of secretory cells along their bellies that exude milk. Marsupial mammals have nipples in folds of their skin, but all placental mammals have full-fledged nipples. Nipples originate in a paired epithelial thickening, which is present in all placental mammalian embryos. Known as the "milk line," this thickening runs lengthwise along the under surface of the body from the limb bud of what will be the arm to the limb bud of what will be the leg — from armpit to groin.

Nipple numbers are highly variable, both within and between species. The general pattern follows the "one-half rule": there are half the numbers of offspring in a litter as there are mammae to feed them. Humans are a case in point, with generally two mammae, one nipple on each, and one young per litter. Pigs have 10 to 14 nipples and rats 10 to 12. Variation in nipple number within species, including our own, is perhaps more surprising. It is estimated that about one percent of humans have extra nipples and/or mammary tissue. The supernumerary tissue can occur in both males and females and can take many forms, ranging from nipples without glandular tissue, vestiges of glandular tissue without nipples, or even full-fledged breasts. Many people have the extras without even realizing it, mistaking a nipple, say, for a mole; pop/country star Carrie Underwood was reportedly one of the supernumerary one percent, until she went under the knife. Extra breasts were one of the indications of witchery in days gone by, and suspects were searched for the "witches' marks" that were thought to provide nourishment for cats, dogs, or more fantastical creature companions. The pattern of high variability in mamma number is evident in many mammals. For example, about 1.4 percent of rhesus monkeys have an incidence of three or more nipples. Extra mammae have also been reported in baboons, guinea pigs, polar bears, and black bears, and this is by no means an exhaustive list.

Generally speaking, mamma number is constrained, in an evolutionary sense, by costs. Mammae create opportunity for infection and cancer. Furthermore, milk production is a relatively costly resource that is limited by the capacity of the mother to obtain nutrients. In part, offspring number is constrained by the extent to which each young requires energy; the more offspring, the lower the "quality" of care, and the greater the chances of the mother not surviving through a tough winter.

Having a small number of extra mammae, however, is advantageous because offspring are not very good at sharing. Behavioral ecologist Paul Sherman's research into naked mole rats has shown that this strange mammal does not follow the "one-half rule." Instead, these prolific, albeit peculiar, mothers regularly have 20 or more pups and only 12 nipples. Naked mole rats likely break the rules for two proximate reasons. One, there is only one reproductive female in a colony at any time, and she is pampered by the other colony members, who share a large proportion of her genes. She does not have to go out and get food like mothers of other species. In fact, she is like a queen honeybee; she does little but reproduce. Second, naked mole rat siblings are highly related,[1] and thus they are much more cooperative than the siblings of other mammalian species who only share an average of 50 percent of their DNA sequences.

Energetic constraints and sibling rivalry may go some way to explain the optimal number of mammae and nipples for a given species, but they do not explain the variation found. Perhaps plasticity in mamma number is itself advantageous, in the evolutionary sense, in that the capacity to quickly adapt to a changing environment over a few generations by responding with increased numbers of offspring imparts higher reproductive success.

---

1 The naked mole rat is one of two species of mole rat that have the distinction of being the only eusocial mammals known; their unique sociality, including the presence of a single reproductive "queen," is explained as it is in social insects, by a high degree of interrelatedness. Within a naked mole rat colony, individuals can share 80 percent of their DNA.

Given a modern genetics education and unlimited time, Bell could very well have been more successful in developing a breed of twin-bearing, six-nippled sheep. Then again, such a fountain of knowledge may have restricted his imagination, and he might never have conceived the wild notion that sheep could have litters. Certainly, he attributed his discovery of telephone technology to a solid ignorance of electronics: "Any good electrician could have told him that he was attempting the impossible."

One measure of the success of any invention is its adoption and use by society. In this sense, Bell's six-nippled sheep are a definite failure. A lack of interest by others, also known as bad timing, may have been a contributing factor in the failure of the U.S. agricultural authorities to grasp on to Bell's multi-nipple legacy and build on his advances. A lack of interest killed more than one of Bell's ideas. Case in point: Bell developed a hydrofoil, which worked exceedingly well. The United States Navy officials that visited Beinn Bhreagh in 1919 to assess it were most impressed with its functionality and applications. It was, in fact, the fastest ship in the world in 1919 — a record it maintained for a full 12 years. However, the man who wrote on the top of the report "this is an old man's toy: a boat that will not fly," did not share the vision, and the invention was rejected by the navy, despite its potential value as a safe, fast water vehicle. Today, the United States Navy does rely on these old man's toys.

To drive home the point about timing, the idea of Bell's multi-nippled sheep so summarily dismissed by the United States Department of Agriculture in the 1940s was reborn in the 1970s, near Dunedin, New Zealand. George Davis, principal scientist at the AgResearch Invermay Agricultural Research Centre, began research into the feasibility of breeding multi-nippled sheep in an attempt to increase milk yields. This sheep research was a sideline to his primary interest, prolificacy — the propensity to produce multiple offspring. At the time, Davis was researching sheep with the high-prolificacy Booroola gene, which was historically introduced into Australian flocks from Garole (Bengal) sheep in the late

18th century and discovered hundreds of years later, in the mid-1900s. These days, the gene can be transferred into *any* flock through the wonders of biotechnology. The Booroola-gened flocks Davis was working with were regularly giving birth to three to five lambs but did not have enough milk to feed them. Davis and his team had successfully bred a number of sheep "configured like a dairy cow," with 40 percent of their milk coming from the two "extra" mammae. However, Booroola sheep turned out to be not such a great idea for New Zealand. They were too prolific, the lambs were small and less cold tolerant, and the ewes didn't have enough food to go around. As a consequence, funding ran out and there was no industry support; prolific flocks were still a thing of the future — common at the research station but not in New Zealand's fields.

There is one other reason that Bell's multi-nippled sheep breeding was a failure: it was not produced as a result of need. Bell was no farmer; he was not attempting to make a living from sheep breeding. Though it is certain that his experiments were maintained by a desire to "benefit the world," Bell did not need sheep. And farmers did not need Bell's sheep. In fact, because Bell had not paid any attention to wool or meat qualities, his sheep were a flock of worthless scrubs. Legend has it that the butcher "declined to take them even as gifts." If there was a pressing need for sheep with six nipples, Bell's efforts may have been better received.

Six-nippled sheep have seen yet another resurgence in recent times, and this time it might stick, simply because it is driven by necessity. George Davis is once again breeding multi-nippled sheep. The new work is being encouraged by New Zealand farmers who've been increasing the prolificacy of their sheep through the use of prolific breeds like Finn and East Friesian and a gene called Inverdale that is being used in the industry. With the Inverdale gene, ewes have a median of two lambs, and often three. But there is a problem; triplets don't do well with a two-nippled mother. Not only is it an issue of enough milk to go around, but feeding tends to take place in bouts, with the lambs feeding all at once. A ewe is not likely to let a third suckle right after the first two have had their fill. Farmers who have Inverdale flocks are now coming to Davis asking for ewes that have more than two functional nipples.

Triplets have a high mortality, and four nipples would greatly help the situation. Davis is going to try to give it to them. He started the breeding work in 2006 with a notice to farmers that the research station would like to purchase any naturally occurring four-nippled ewes or rams, though, unlike Bell, he only pays market prices. Like Bell, Davis is particularly interested in six-nippled sheep. It turns out that to consistently get four functional mammae, six-nippled sheep give the breeding program a real "kickstart." If Bell's experimental results are any indication, the program will be a resounding success, and there is every reason to believe that multi-nippled sheep will become a fixture on the future pastures of New Zealand and beyond. It appears that Bell did indeed give birth to an idea of merit. It just had to wait for necessity to come along to nurture it.

Inventions are like mutations: environment has a great deal to do with their success.

ELIHU THOMSON'S
QUARTZ TELESCOPE
MIRROR

1853–1937

This undated photograph shows Thomson with a fused quartz mirror. The caption on the back reads, "The mirror shown in photograph is 22" diameter. A 60" diameter mirror is now being completed. Subsequently a 200" diameter mirror is to be made."

Elihu Thomson has 700 patents to his credit — all of them useful. This makes him a contender for the title of greatest inventor ever.

Thomson's many engineering feats put the whiz in our washing machines and the zip in our car engines. His first successful invention, a cream separator, is the forerunner of today's centrifuge. He was instrumental in the development of electric motors; his career in that field was marked by the invention of the three-coil dynamo — the first three-phase generator. Decades before Hertz and Marconi, Thomson described electric waves and recognized the potential to use the movement of electricity through space to communicate. Electric lamps, the electric meter, X-ray machines, musical instruments, lightning rods, electric furnaces, streetcars, automobile engines and other car parts, and the use of inert gases in deep sea diving are among his contributions. He invented the grounding system that keeps electric conduction safe. Electrical welding was entirely his idea; he developed the means whereby metal could be fused together without solder. Metal tubing and car parts were some early applications of this technology. He also made significant scientific advances in the earth sciences, chemistry, medicine, and astronomy. And he was involved in the development of two major corporations, the General Electric Company being the better known.

Thomson understood the roles of finance and industrial research, and, unlike some of the others in this book, he was not a prima donna. Rather, he enjoyed working in teams, leaving others to their strengths and capitalizing on cooperation for mutual benefit. He was also wise and able to protect himself to some degree from the greed of others. For example, he insisted in writing a clause into the contract with his first company, the American Electric Company, which stated that should the company not perform its duties fully, all of Thomson's patents would be returned to him and the company dissolved. The shape of the current infrastructure for supporting technology in the public and private sector can be, in part, attributed to him.

Thomson was known as "the Professor" — the one who could

be called upon to assist in any problem, seemingly in any field. He was the type that could knowledgeably speak in depth on any subject at a dinner party. Kindhearted and quick-witted, he made life-long friendships readily. The anecdote that follows reveals something of his character. At the Electrical Congress at the Chicago World's Fair, the likes of Tesla, Edison, and Thomson gave talks and demonstrations. At the final banquet dinner, Edison was asked to speak. Thomas Edison abhorred public speaking and resisted, but the crowd continued to insist, making for a highly awkward moment. Thomson rescued him by standing up, drawing attention to himself and announcing, "Well, if Tom will not speak, Tom's son will have to."

The strangest thing about Thomson was probably his first name. His biographer said of him: "As a biographical subject the Professor . . . bled little, and his hurts were the universal ones of personal bereavement that were not news. His life moved smoothly through its eighty-three years upon a course appointed by the gods themselves. He struggled with no rivals, created no hatred anywhere. He bequeathed to his biographer a beautiful problem in dramatizing the conventionally undramatic."

Everyone respected Thomson, except perhaps charlatans, pushers of pseudoscience, and the religious, for whom he had a singular intolerance. His mother-in-law was a strict Congregationalist who attempted to convert him. He did not concede and retorted, "Church was a place where you had to listen to a lot of things you disagreed with, told you by a man you couldn't talk back to." At the Chicago World's Fair, he led a protest meeting aimed at the exposition's administration, who had allowed the quackery to present their products side by side with true engineers.

Thomson's inventions were generally not major breakthroughs in our understanding of the universe, but rather they were practical solutions to practical problems, carefully thought through from theory to practice and based on logical, scientific principles. He was a wide-ranging genius who began inventing and discovering as a teenager and never stopped. While Edison tinkered, believing his failures were steps toward a solution, Thomson's inventions tended to be near right the first time, the bugs already worked out in his mind by solid reasoning.

So strong is the empiricism in Thomson's veins that he is regarded first as a scientist and second as an inventor. His scientific curiosity appeared to have few boundaries, and he delved into others' questions as readily as his own. One example of this tendency surrounds the controversy of the safety of X-rays. In 1896, Nikola Tesla was publicly claiming that X-rays were harmless, being of the nature of light. Thomson and others were terribly concerned because Charles Dally, who was employed by Thomas Edison to make X-ray tubes, fell quite ill as a result of his work.[1] Thomson decided he needed to settle the matter once and for all to prevent further victims (and was asked to do so by researchers who knew him). He decided which part of his body he was best able to part with, settling on the tip of his left baby finger. Using a lead shield to cover the rest of his hand, he exposed his pinkie to an X-ray tube for 20 minutes. The exposed portion was "burned." He further experimented with the tip of another finger, using various materials as a shield — proving unequivocally that X-rays are dangerous. The results were published widely, and the scientific community took due notice. For the rest of his life, the two sacrificial fingers were scarred and stiff.

Among the creative, eccentric genius of his great inventor peers, Elihu Thomson reminds us that creativity can be sure-footed. Or, put another way, he demonstrates that empiricism at its best can match creativity in its output, or at least come pretty close to it.

There are failures in Thomson's record, though they are few and far between. His first two inventions earned a weak response: a fastener for streetcar rails, which was used only experimentally, and a pneumatic relay sounder for sending telegraphs, which also did not catch on. However, Thomson's attempt to develop a giant mirror

---

[1] Edison immediately discontinued work with the tubes. Charles Dally died in 1902 from the consequences of X-ray exposure.

made of quartz for a telescope is one solid example of failure in a near perfect career.

In Thomson's day, interest in astronomy was strong, and his own interest was roused early. At 4 a.m. on November 15, 1867, 14-year-old Elihu and his siblings were woken up by their mother and ushered out of the house to watch a meteor display until dawn. Later in life, Thomson attributed his love for the sky to this experience. As an adult, Thomson continued with this tradition of nocturnal expeditions and traveled thousands of miles to witness solar eclipses. An interest in optics was also present at an early age. While still a boy, Elihu made: magnifying lenses from the bottoms of old pharmacy jars, a microscope from melted glass thread, and a camera, of sorts. He graduated to the making of fine lenses and in 1878 published a paper describing a simple method to shape the surface of an astronomical mirror. In 1900, he built a 10-inch (25 cm) telescope for his own private use. In Thomson's time, large telescopes were coming into being, and astronomers were busy looking at the planets, the Sun, and the stars. Mars held a particular fascination, and observations of canal-like lines on this planet's surface led to a scientific debate over their possible significance to the question of a Martian race of humanoids. True to form, Thomson dismissed the idea in favor of a more parsimonious explanation involving patterns of vegetation.

The universal desire to see farther and farther into space drove the need for larger and larger telescopes. The fundamental function of a telescope is to gather light and focus it. The first telescopes were made with refractive lenses — a series of glass pieces shaped to gather light entering from space and focus it. Size has always mattered when it comes to telescopes, as the larger the area that is gathering light, the farther into space one can see and the better the detail of the pictures from closer celestial bodies. Carefully positioned mirrors could do the same job of gathering and focusing light, with the advantage that only the surface of the mirror is interacting with the light, rather than the light passing through a thickness of glass. This results in less opportunity for distortion of the images. Today, all large telescopes use mirrors.

In Thomson's day, regular silica glass was the material used for mirrors and lenses. But problems were arising as the size of mirrors and lenses was pushed to the limit. The large plates of glass were expanding with heat from the Sun, warping the mirror's surface and shattering the precious images from space. Thomson had the answer. Starting near the turn of the century, he began experimenting with fused quartz as a material for these mirrors. In his own words, he compares fused quartz with regular glass:

> I found that by instantaneous application of a moderate heat or a small flame on the back of the glass slab, the image went immediately all to pieces, as we may say; that is, it scattered; the definition was gone. A similar treatment of the quartz slab showed very little change, and not until the back of the quartz had become quite hot was there a semblance of the disturbance such as occurred with the glass. This experiment, modest as it was, convinced me that there was one material suitable for the making of astronomical reflectors that would avoid many of the difficulties of construction and operation inherent with the glass mirror telescopes.

Fused quartz is quartz rock that has been melted at the very high temperature of 3,632°F (2,000°C) to destroy its crystalline structure. Greg Durocher from United States Geological Survey explains it thus, "Fused quartz and silica have the same formula, but [are] refined from different feed-stock."[2] Fused quartz is very pure and transparent and has a lower coefficient of thermal expansion (CTE), so it does not expand as much as glass does when heated. Fused quartz also has a higher transmission of ultraviolet wavelengths, which is useful for detecting this portion of the light spectrum from space, with particular applications for understanding star forma-

---

2 Some of the terminology is confusing. As Greg Durocher clarifies, "Both fused quartz and fused silica are examples of vitreous silica — quartz glass — a clear and colorless glass made only from SiO_2. The term 'quartz glass' might be going out of style, but it still often implies a fused silica of optical quality."

tion. (Young stars produce a lot of uv rays.) All in all, it is a superior material for telescope mirrors. Not only did Thomson recognize the superior qualities of fused quartz, but it was only because he'd invented an electric furnace, which produced the heat necessary to melt the rock, that the use of this material for telescopes was possible at all.

When the prominent astrophysicist George E. Hale set up his second observatory, Mount Wilson Observatory, near Pasadena, California, he had his colleague Elihu Thomson supply him with quartz mirrors. But it did not work out. The surface of the mirror was full of microscopic air bubbles that made it impossible to accurately interpret the light collected from space. Over 25 years later, Hale was setting up another observatory, on Mount Palomar, California. He wanted a 200-inch telescope, which would be the largest in the world at the time. For comparison, the Hubble Space Telescope, launched in 1990, is this size, and it is still providing us with information about the expanding universe. Hale wanted fused quartz for the mirror because of its low CTE and the lower likelihood of it warping. Low CTE is also a desirable property because the mirror does not heat up so readily in the polishing process. Since testing of the mirror has to wait until it has cooled down, quartz mirrors can be polished and finished in much less time than a comparable regular glass mirror. In 1928, Thomson's laboratory agreed to attempt to make the mirror for the Mount Wilson Observatory, at cost.

Then 75 years old, Thomson led the experiments. To get around the air bubble problem, Thomson and his team attempted to melt a clear layer of quartz onto the quartz slab — "enamel" is what Thomson called it — which would provide a perfectly smooth surface. The experiments continued for years. The melted quartz was sprayed on in thin coats, and the layer built up slowly. To accomplish this, a blowtorch was invented to provide a flame hot enough to melt the quartz. Thomson was confident of success and wrote to Hale in 1929: "We now have no doubts of the possibilities of the construction of 200-inch mirrors." But success eluded them.

Though Thomson remained confident that given sufficient time he would find a way, his time ran out. Or more importantly, the money ran out. Letters from Hale in 1931 speak repeatedly of staggering expense, complain diplomatically about the size of the bills, and suggest nervousness from the financial backers of the project, the Rockefeller General Education Board. Bills exceeded resources in May of that year, and Hale broke the bad news by letter. They tried to obtain funding for some time afterward, but in 1935 officially called it a day. The expense of the work had totaled $600,000, and in the end no 200-inch mirror was ever attempted from fused quartz. Thomson was stopped at 60 inches, a far cry from his original goal. The 200-inch mirror for Mount Palamar was made with Pyrex glass instead, and it supplied astronomers with a fantastic tool for exploring space.

It is said that Elihu Thomson took the disappointment stoically and continued unchecked with his other work. He had a deep understanding that there was much more to successful invention than a great invention. He once told a fellow inventor, "There is a saying in the good book that the bread which you cast upon the waters returns. But it's very often somebody else's bread. And the man who invents a thing ahead of time is usually forgotten by the latecomers, who get the credit." In the case of the quartz mirror, lack of money, and time, stood in the way of progress.

Thomson was not incorrect about fused quartz being an appropriate material for the task. Fused quartz is still considered one of the finest materials available for telescopes and has been used to make large, important telescopes such as that on NASA and Stanford University's Gravity Probe B satellite. Fused quartz is manufactured by GE, among others around the world, and is readily available commercially for anyone who wants it. Today, new materials have also come into use, including Zerodur, a ceramic glass.

Today, the world's largest single piece of glass, the Subaru Telescope in Hawaii, is just over 27 feet (8.2 m) in diameter. It is 8 inches (20 cm) thick and weighs 25 tons (22.8 t). It is made from Corning Code 7972 Ultra Low Expansion (ULE) glass coated with aluminum. Corning 7972 glass is a titanium silicate. It has a thermal expansion coefficient (CTE) of 0 ± 30 ppb/°C, compared to a CTE in the range of 5.60×10^{-7}/°C for some forms of fused quartz. For comparison, Zerodur has a CTE of about 3×10^{-6}/°C. In layperson's terms, Corning 7972 has a CTE in the region of about 20 to 100 times lower than fused quartz, which has a CTE about five times lower than Zerodur. The Subaru mirror is so large that it does distort, but this is physically readjusted by "computer-controlled support that has 261 small motors with screens and sensors — actuators — that push and pull the mirror to the right shape as needed." With the Subaru, and other huge telescopes, astronomers are looking out into the far reaches of time and space at the first stars and young galaxies. Elihu Thomson would have been thrilled by these achievements, even though the sands of time have precluded him from being among those to see it.

There are at least five dimensions in invention: the three dimensions of space, plus time and money.

THE
MODERN
ERA

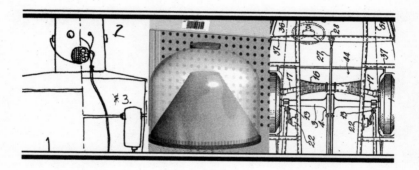

PAY ATTENTION TO DETAILS

DANNY HILLIS'S

PAINT CAN ROBOT

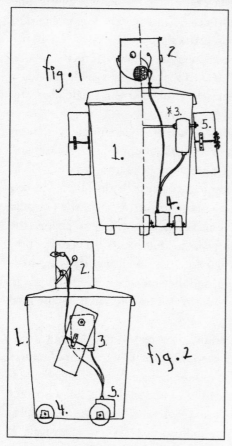

1956–

Artist interpretation of Danny Hillis's robot made from paint cans. (See endnote.)

Each year, the MIT computer science department picks a few theses out of its graduate student harvest for publication. In 1985, Danny Hillis's landmark development in computer hardware was one of them. His Connection Machine was revolutionary. It was a "computing engine" in which processing was accomplished concurrently through a massive number of simple processors, with a resulting substantial increase in speed and efficiency. Simply put, processing in the Connection Machine occurs in parallel, as opposed to the sequential processing of a linear processor. In a basic scenario, one instruction is executed at a time in linear processing, whereas in parallel processing, numerous instructions are executed simultaneously, which saves time. If a computer is a restaurant and the processor is the kitchen, then Hillis's parallel computer is like having a team of cooks to prepare the meals, instead of a single chef; the food gets served a lot faster.

Despite its success, Hillis says the parallel computer was ahead of its time. While Hillis foresaw a market for it from the outset and set up the company Thinking Machines to sell it, it did not do as well as expected, and the company eventually dissolved. He says that there was so much emphasis and investment in the personal computer back then that the market was unwilling to change the software. It took more than 20 years, but parallel processors now predominate modern personal computer architecture (i.e., dual-core or quad-core processors), and they have been used for years in "supercomputers" for a wide variety of research and commercial applications.

In contrast, the RAID (redundant array of independent disks) hard-drive system Hillis developed for the Connection Machine, which uses multiple hard drives simultaneously, was a big commercial success from the outset. Hillis, however, almost did not patent it. He thought that such a clunky system, which uses mechanical motors, was surely technology on its way out. Boy was he wrong. It just goes to show that inventors "live in the future."

Such forward thinking is evident in Hillis's other activities. In fact, he's a founding member of the Long Now Foundation, which

promotes long-term thinking. The organization is bent on adjust-
ing our idea of time and stretching it well beyond our own puny
experience to the vast, and more realistic, timescale of the Earth.
The concept was Hillis's:

> The future has been shrinking by one year per year for my entire life.
> I think it is time for us to start a long-term project that gets people
> thinking past the mental barrier of an ever-shortening future. I would
> like to propose a large (think Stonehenge) mechanical clock, powered
> by seasonal temperature changes. It ticks once a year, bongs once a
> century, and the cuckoo comes out every millennium.

The first prototype of a 10,000-year clock-of-the-long-now,
completed by Hillis in 01999, is housed in the Museum of Science,
London. Another clock is in progress that measures the same time
frame but works on the basis of a sun and planet model — a modern
astrolabe. The workings of the clock are technologically unique; it
has a precision equal to one day in 20,000 years, and it self-corrects
daily according to the noon sun. Hillis has several patents from this
work. The final design for the clock will be very large. It will eventually rest on a mountain in Nevada with a white limestone face,
which has been purchased by the foundation for this purpose.

Hillis has also turned his attention to the Internet, with the creation of a new project, www.freebase.com. The website is described
as a "social database about the things you know and love," and it
was publicly launched in beta phase in 2007. This public database
is to Wikipedia what the Connection Machine is to the von
Neumann computer. While Wikipedia's database has a hierarchi-
cal structure, with each bit of information having its own place as
a subset of a bigger category, Freebase is flexible. Each collection of
information is thought of as a hub, with each hub connected to
many others. Bits of information can belong to however many hubs
people decide they do. To this Internet grazer, who is a gentle her-
bivore, so to speak, rather than an autotroph or carnivore of Web
technology, Freebase seems to be powerful stuff. It has a graphic feel

to it and gives the impression of a leap forward in how knowledge is stored and used. Freebase encourages users to create novel applications that use the database and provides some tools to do so. With Freebase, Hillis is taking his ideas of a brain machine one step further — he seems to be aiming for a global brain with this invention that is analogous to the prefrontal lobe or hippocampus or perhaps an altogether unique kind of neural network. Hillis shared something along this line of thought with the web magazine *Edge*:

> One might suppose that, with all its zillions of transistors and billions of human minds, the world brain would be thinking some pretty profound thoughts. There is little evidence that this is so. . . . For the most part, the Internet knows no more about the information it handles than the telephone system knows about the conversations that take place over its lines. Most of those zillions of transistors are either doing something very trivial or nothing at all, and most of those billions of human minds are doing their own thing. . . . The hierarchical teams that built these [pyramids] were an extension of the pharaoh's body, the pyramid a dramatic demonstration of his power to coordinate the efforts of many. Pyramid builders had to keep their direct reports within shouting distance, but electronic communication has allowed us to extend our virtual bodies, literally corporations, to a global scale. The Internet has even allowed such composite action to organize itself around an established goal, without the pharaoh. The Wikipedia is our Great Pyramid. . . . What is still missing is the ability for a group of people (or people and machines) to make collective decisions with intelligence greater than the individual.

There are other works in Hillis's oeuvre, and he holds 50 United States patents to date, ranging from simple devices like a laptop computer desk stand to more sophisticated lens systems, sensors, and an apparatus to maintain eye contact during teleconferencing. He has two patents for methods of masking speech, with applications for anyone who wants to speak on the phone with privacy. There's also an atmospheric spectrometer for use by planetary satellites, a method for locating and identifying underground structures, a control wand, a robot for military use on the field, a 360-degree awareness system for drivers in a car (instead of just rear and side-view mirrors), and more.

Many of Hillis's inventions, including the parallel computer, have been part of a broad interest in artificial intelligence (AI). In the introductory pages of his PhD thesis, *The Connection Machine*, he compares the human brain to the computer, pointing out the ostensibly disparate attributes of the brain, its relatively slow operating speed coupled with its ability to outthink a machine with ease:

> What the human mind does almost effortlessly would take the fastest existing computers many days. . . . As near as we can tell, the human brain has about 1010 neurons, each capable of switching no more than a thousand times a second. So the brain should be capable of about 1013 switching events per second. A modern digital computer, by contrast, may have as many as 109 transistors, each capable of switching as often as 109 times per second. So the total switching speed should be as high as 1018 events per second, or 10,000 times greater than the brain. Thus the sheer computational power of the computer should be much greater than that of the human. Yet we know the reality to be just the reverse. Where did the calculation go wrong?

The excitement and frustration in Hillis's words about the potential and challenges of AI is almost tangible. His fascination was nurtured in tandem with efforts made in inventing robotics. Hillis's first robot was a spectacular piece of engineering. Standing several feet tall, the humanoid had a head made from an empty 1-gallon paint can sitting on a trash-can body. Its four limbs were made from 1-quart paint cans. The Christmas tree–bulb eyes lit up, and it had a speaker for a mouth, wired for sound. Its arms moved via the use of a rotisserie motor, which was rescued from a discarded barbecue. Hillis built this robot with the explicit purpose of having a companion to walk with him, or at least bring him orange juice in the morning, and he was exceedingly disappointed when the invention failed. Like all inventors, Hillis had a vision, and he built that vision. But as is common in the inventor's process, he had neglected to visualize a key detail required to make the invention functional. In this case, he'd forgotten to give his robot a brain. Danny Hillis was only nine years old at the time, so his spectacular failure is probably forgivable. Besides, in the broader context, this particular paint-can robot served a very useful purpose. For in that pivotal moment when the robot did not do his bidding, Danny realized the importance of the "machine brain," and his lifelong interest in computers began. We have this paint-can robot to thank for that interest, and the substantial benefits to society it later produced.

At a much later date, Hillis did make a working robot, fulfilling his childhood dream — and, in a sense, repeating his childhood failure. While vice president of research and development for Walt Disney Imagineering from 1996 to 2000, Hillis wanted to create a unique attraction for theme park visitors, something sensational that would provide thrills and entertainment in keeping with Disney's philosophy. Hillis wanted to build a robot dinosaur that wandered around the park, interacting with the public as it did so. The executives were not convinced this was such a good idea. Compared to riding a roller coaster, a wandering metal dinosaur sounded a little low on the thrills side of the equation. But Hillis was given a green light to build a prototype.

When the day arrived to reveal his masterpiece machine, Hillis knew he had a challenge ahead of him. He needed to prove to his colleagues that being approached by a metal dinosaur was indeed a thrill worth paying for, one that would evoke wonder and awe. Hillis planned the unveiling of his robot carefully. In a warehouse, he hung a large, dark curtain as a backdrop. He strategically positioned a large wooden crate in front of it, plastered with "Danger" and "Do Not Open" stickers. At a respectable distance, he arranged a semicircle of chairs for those he needed to impress. With everyone seated, the moment of truth had come. The door to the crate was opened. And then . . . the room was filled with the thunder of giant footsteps, the giant curtain came down, and a 14-foot tall triceratops rushed from behind it, heading straight for the audience. The Disney execs were impressed all right — they jumped out of their seats and ran for cover, thrilled out of their wits. In Hillis's words, he had "oversolved the problem." The dinosaur was declared a little *too* thrilling for public consumption. Even though it was a technological marvel — 6 tons (5.5 t) of metal that could move as fast as a running human but was gentle enough to step on a light-bulb without breaking it — the dinosaur was simply too realistic. In a sense, Hillis had failed again.

Hillis has talked about how and why inventors fail — he says, "Inventors don't think of failures, they are just things that are not appreciated yet." More seriously, he thinks it is typical for inventors to have blind spots — as a group, inventors can be pretty excitable. During the invention process, Hillis explains that he visualizes an idea. He says that with experience he's gotten skeptical of his enthusiasm and recognizes that his vision of an invention is just an illusion created in his own mind, one that is likely faulty and will need adjustment if it is to work.

Of course there is nothing wrong with Hillis's robot ideas. Robots to deliver orange juice are now a reality. As of 2006, one Hong Kong restaurant "employs" three of them to take orders and deliver the food and drinks. Customers love it, and the robots have definitely helped business. Though, apparently, they are pretty inefficient — so much so that the restaurant actually requires more staff than a normal restaurant. Caring for robots is one of the core duties of the human staff.

It seems that we are living at a critical period in the evolution of the humanoid robot — when the use of intelligent machines is experiencing rapid change. A robot chef now exists that, given the raw ingredients, can produce more than 160 classic Chinese dishes. Honda has developed ASIMO, a humanoid robot that walks like a human, talks like a human, kicks a soccer ball like a human, and brings people orange juice. It can also recognize faces and perform commands spoken to it.

MIT is actively researching the production of robots with social skills, pushing the envelope closer and closer toward the development of a humanlike intelligence that includes social interaction and common sense or, perhaps even more scarily, an un-humanlike intelligence. That robotics has the potential to reach I, Robot proportions is hinted at by the very real consideration by the British government that human rights for robots could become an issue within the next 50 years. Robots are also frequently mentioned in the answers to the 2009 Edge Annual Question "What game-changing scientific ideas and developments do you expect to live to see?" Just how close to the complexity of human emotion and experience a machine can get is a matter of debate, but Hillis's paint-can robot now has some sophisticated relatives.

Hillis's Disney dinosaur has also been spawned. Disney released a robotic dinosaur into one of their theme parks, albeit a toned-down version. "Lucky" made his debut in 2003. This smiling 9-foot (2.7 m) tall, 12-foot (3.7 m) long two-footed dinosaur with bright green skin and a toothy smile was released into California Adventure to interact with the public. Lucky sounds rather endearing; he has a squishy nose, makes cute little noises, pulls a flower cart, sneezes, gives autographs, and can even steal a visitor's hat. He has a face that looks like a baby sauropod and the upright tail and bipedal gait of a T. rex. The live interaction is particularly pleasing to visitors, though it isn't completely the result of robotics — an "invisible" operator controls Lucky's movements. Lucky is the invention of another MIT graduate, Akhil Madhani; Danny Hillis's contribution to the concept is not part of the official story.

Robotic dinosaurs, in addition to those featured in movies, are accessible to the public in other ways. Monster Robots Inc. has a 40-foot (12 m) tall, 30-ton (27 t) dinosaur machine called Robosaurus that is available for hire for events, photoshoots, etc. The creators of this mechanical beast recommend it for, among other things, taste-test promotions. Robosaurus has been designed to lift a 55-gallon drum up to its mouth to "drink" its contents. It can then turn and crush the competition in a thrilling display of strength and brand loyalty.

Another inventor, Caleb Chung, has fashioned a smaller dinosaur robot, a baby sauropod called Pleo. Currently retailing for $369, this robot pet (or artificial life-form, as its creator calls it) snores, plays, and responds to human behavior. Its personality is developed by its interaction with its owner, so a Pleo can be "raised" to be a sweet, cuddly personality or a vitriolic, annoyed one.

Despite the progress made in robotics, in and outside Hillis's labs, things haven't changed much for him. He's still thinking into the future, he's just traded global brains for robot brains. And instead of rummaging in his parents' garage for paint cans and barbecue parts, his company headquarters has a warehouse scrap pile that Hillis jokes is "full of really useful junk, like electron microscopes and spectrometers."

Regardless of the scope of the invention, failure is simply part of the process. In the words of Hillis himself, "zillions of prototypes" are normal.

BE AGREEABLE

J.WALTER CHRISTIE'S
FLYING
TANK
1865-1944

Christie's flying tank came to fruition in Russia and Japan. This Antonov A40 is shown in a test flight.

J. Walter Christie got off to a good start. At 39, he was racing in the first international car races in the United States, the Vanderbilt Cup, as a serious contender.[1] He would soon tie the world record for gasoline cars on the circular Morris Park one-mile track, with a time of 51.5 seconds, and would go on to break the American record for speed on a straightaway. In 1907, he would also be the first American to race in the French Grand Prix. At that point, not only was Christie a world-famous driver, but he was also known for his cars, which were of his own design and build. His unique 50-horse-power front-wheel-drive car was completed in 1906, sold for $6,500, and raced in the Vanderbilt races. Shortly thereafter, however, the company that he'd set up to make and sell this vehicle, the Direct Action Motor Car Company, failed. Christie's exciting racing career also came to a quick end when he crashed, in Pittsburgh in 1907, going 70 miles per hour (113 km/h).

After recovering from broken bones and abdominal injuries, Christie's next plan was to design front-wheel-drive taxis for New York City, but before that plan had a chance to transpire, he saw a notice put out by the Los Angeles Fire Department. The request was for bids for the manufacture and supply of tractors to pull their fire wagons. Replacing the teams of horses with a vehicle was a pop-ular strategy at the time, as it modernized the firefighting equip-ment without the extra cost of replacing the chemical and hose truck as well.

Christie needed the money, and his business, then called the Front Driver Motor Company and located in Hoboken, New Jersey, put forward one of four bids for the LAFD job in 1912. His was the lowest and was accepted. Fate had driven him one step away from cars into bigger, heavier machinery. The "Christie," as the fire trac-tor came to be known (like the inventor's cars before) "was an odd looking affair with its 80 horsepower, water-cooled motor sitting crossways on the front of its frame; the bare flywheel twirling away

---

1 These races were held between 1904 and 1910. For more info see www.vanderbiltcupraces.com.

unprotected on the right side, and the driver perched high behind a vertical steering wheel above and behind the motor." Four tractors were sold to the LAFD for $4,500 each. In the years that followed, the LAFD was happy enough to purchase three more of them. The New York Fire Department subsequently bought vehicles from Christie too. In all, Christie sold between 600 and 800 fire engine tractors — a brisk trade. However, when the LAFD approached Christie in 1917 with yet another order, they were turned away. Christie had moved on to even bigger things: tanks.

In 1914, Britain and France were engaged in trench warfare with the central powers of Germany and Austria-Hungary along the 475 miles (764 km) of the Western Front. Infantry manned the trenches, the cavalry was waiting in the wings, and the span of ground between the sides was deemed no-man's-land. It was a stalemate, of sorts. In 1915, a British officer, Lt. Col. Ernest D. Swinton, got an idea while watching a caterpillar tractor in France: why not use a similar but armored vehicle to cross no-man's-land and deliver firepower? He presented his idea to a committee back home, and one man who heard the idea was impressed. He was a navy man and first lord of the admiralty; his name was Winston Churchill.

Churchill spearheaded the experimental phases of a "landship" in 1915. The program was kept secret, and early models shipped to France from England were transported in crates labeled "water tanks" for Russian use. This ruse is the origin of the term "tank," now used for what was originally, literally, an armored agricultural vehicle. The British first used tanks in combat in 1916. By 1918, they had 600 in operation in the Battle of Amiens. The Germans were quick to initiate their own armored vehicle program and had built 15 units of a gigantic 30-ton (27 t) machine by the end of the war, which were used in later battles. In three short years, the tank had gone from a caterpillar tractor to a fully operational armored combat vehicle. These early vehicles were not, however, particularly reliable. Problems were multifold, including mechanical failures and the need for protection. Though armored, the vehicles were still vulnerable to attack. Nonetheless, the tank had arrived.

When the United States entered the war in 1917, it too had decided to make tanks of its own. Walter Christie jumped at the chance to turn his vehicle designs to warfare. While it may have been somewhat opportunistic, Christie did have a genuine interest in military engineering, and thereafter he devoted himself to tank design regardless of the obstacles he encountered. He had dabbled in military engineering in the past, with a patent for a ring-turning lathe granted to him in 1904.[2] The lathe was designed to manufacture, with greater accuracy, the circular tracks on which a gun-turret turns. In 1916, Christie made a prototype for a four-wheeled gun carriage. Then, for more than 30 years, he made prototype tank after prototype tank, and engaged in battle with his countrymen in the United States military to persuade, and later to coerce, them to support his work.

In the early 1920s, Christie developed three versions of an amphibious tank. The concept was not his originally, but Christie brought to it a genius for making vehicles that did remarkable things. This genius is illustrated by a description by General S.D. Rochenbach, United States First World War tank leader, who witnessed a demonstration in the Hudson River of one of these amphibious tanks in 1922:

> Its climbing ability was something remarkable, and it continued to ascend for 100 feet, where the earth ended against the precipice. It then descended the bank much easier than the spectators, who were slipping and sliding, reached the road, turned abruptly to the left, went along the river for some 30 feet, and then descended a 6-foot stone wall into the Hudson River. It crossed the Hudson under its own power, but instead of the driver directing the machine to its designated landing place on the east shore, he headed south for the same and on getting across the river, faced a sheer stone precipice. . . . It was the most remarkable performance that I have ever seen or heard of. . . .

2 US Patent Number 759,592.

Despite its apparent success, and the impact its demonstrations had on the public, the United States government did not buy this, or any, amphibious prototype. Eventually the vehicles were destroyed. Christie did, however, manage to sell the plans to the Japanese, so his efforts were not a complete waste.

Christie worked on land tank designs too. His first tank, the M1919, was constructed with a truck chassis. It had front-wheel drive, a convertible track/wheel system, and a unique suspension system. These elements, particularly the suspension system with independently sprung wheels, created a vehicle that had much greater vertical movement of its wheels. This was one unique feature of Christie's designs. Christie believed, and some in the army like General Patton agreed, that what was needed was a tank that had wheels, to drive efficiently and quickly on roads, that could be converted to tracks, for maneuvering over difficult terrain. There were significant problems with the M1919, however, and it was mechanically unreliable.

Christie continued to work out the bugs, and in 1928 he unveiled a new model, the M1928. As in the past, Christie revealed his work to the world with public demonstrations and fanfare. The demonstration caught the eye of cavalryman C.C. Benson, who called the tank "the M1940" in an article for the military rag *Army Ordnance* because it was "easily ten years ahead of its time." This machine, weighing 8.6 tons (7.8 t) without armament,[3] could travel 70 miles per hour (113 km/h) on wheels and 42 miles per hour (68 km/h) on tracks. The chief of ordnance recommended $250,000 be set aside to purchase several of them from Christie, for testing. Congress agreed, but the order was later revoked when the position of chief of ordnance changed hands.[4]

---

3 After his first couple of tries, Christie built all of his prototypes without any armament. Critics in the United States military did not like this, as it made it difficult to predict how a tank would actually perform with the added components; obviously, tanks needed guns or there was not much point in having them, no matter how fast they could go.

4 $62,000 was used to purchase one single test M1928, and the rest reverted back to the Treasury.

The Russians saw Benson's article, came and had a look at Christie's design, and decided it was just what they needed as the basis for a light tank to round out their tank development program. They signed a contract with Christie and bought two M1930s — updates of the M1928 model — in 1930. Christie shipped the wares to the Soviet Union as "agricultural tractors." United States officials later learned of the trick and confronted him, but he denied any wrongdoing. Christie's suspension, and other elements that produced such a fast combat vehicle, were fundamental in the subsequent Russian BT tank series. Christie actually signed a contract with the Russian military agreeing not to "transfer or otherwise dispose of his patent rights or interest in said tanks to any third party" and to provide all blueprints, parts, and instructions as required. He did not maintain the exclusive clause of the contract and sold one tank to the Polish, considered the enemy of Russia, a few months later.[5] He also sold one tank to the United States Ordnance Department, though the deal was never finalized.

There were 10 more prototype tanks built by Christie, the last in 1942, when the inventor was 77 years old. The United States government never made a bulk purchase of Christie tanks and never did what Christie claims he really wanted them to do: "Give [me] the money and let's see what kind of machine [I] can turn out." In the United States, he had to scrape and claw to get his own government to purchase a few models; seven M1931s were purchased in 1931 and 1932. These were the last of the Christie tanks sold in the United States with Christie suspension; wheel/track convertibles were abandoned by the United States altogether by 1935.

Christie's designs, however, were quite influential in other parts of the world. In the Soviet Union, large-scale maneuvers were being run with BT tanks, which were based on the Christie design. The Red Army used the best features of the Christie in their tank

---

5 This deal fell through, however. Christie did not have the tank ready on time, and he eventually paid back their deposit.

program, culminating in their T34, which was specifically designed "for deep operations" (breaking through the enemy's defenses and swiftly moving to an operational depth). The T34 has been called "the epitome of creative design" and "a masterpiece of tank crafting." The British acquired a Christie M1930 and used parts of it in a redesign that was to become the A13 Cruiser. By comparison, the prevailing United States tank during the Second World War, the M4 Sherman, was apparently not up to par.

There are several reasons for the mediocre success Christie experienced at home. For one, the United States did not have an organized tank program nor the budget that goes with one. (That said, George Hoffman, who has written more than anyone about the "Christie tank controversy," has said that the United States government paid Christie $739,240 for his tanks between 1916 and 1921; by all accounts, Christie had financial troubles during his entire tank-building career, so the funds were all spent on development.) The lines of authority and decision-making were blurred regarding tank development, and the army, cavalry, and ordnance departments could not agree. Christie had some strong supporters — one official enthusiastically referred to the M1928 as "a wildcat." There were also many naysayers; the same tank model was declared "structurally unsound and only good for flash demonstrations." Changes in command in the middle of negotiations did not help matters, and there was also Christie himself, who everyone agreed was difficult. General Patton, a fan of the Christie tanks, called Christie's less pleasant side "histrionic inclinations," and Colonel Icks said Christie "goddamned everyone and everything."

Christie had proven himself capable of duplicity. He did not play by others' rules, and he was stubborn about his ideas, as are many great inventors. To his credit, it must have been exceedingly frustrating for him to be a United States civilian designing tanks when the United States government failed to see the value that other countries saw in his designs. Perhaps the United States did not support him more readily because he was not "one of them"; he was civilian, not military, adding a psychological twist to the "not invented here" syndrome of big business versus the independent inventor. However, it might also be that Christie's hostile, grating personality made cooperation impossible by all except those who had at least an ocean to separate them from the man and his troublemaking.

In 1931, when the Ordnance Department tried to buy an M1930, Christie refused to fill the order unless he was also given an order for additional tanks. The tank remained unsold. In 1933, $200,000 was set aside for tanks in the federal budget, and Ordnance wanted Christie to bid to supply those tanks; however, he refused, saying "the specifications as prepared do not conform to the advanced art in the construction of tanks and contain requirements which this company could not and does not desire to comply with in view of the improvements already tested by it." In other words, he didn't want to comply with his client's needs; he wanted to build Christie tanks that met his own ideals of tank design, whatever the United States military thought.[6] Christie wanted the military to buy the M1928 already designed, as is. When they refused, he parked the vehicle in the courtyard of the State, War, and Navy Building! The $200,000 went unspent.

In addition to Christie's abrasiveness, there were other aspects of his personality that affected his inventions' success. He was impractical as well as stubborn. While he understood some of the military's requirements for combat vehicles — like the problems of traveling from one place to another, which his convertible track/wheel system and emphasis on speed addressed[7] — he ignored others, such as armor. Christie's prototypes were fast, in part because they were lacking in adequate armor thickness. Other aspects of use in field situations were also ignored or de-prioritized, like adequate room for a crew and guns.

---

6 The other side of the story is that the United States military could not decide what it needed and floundered about with internal arguments and no clear direction for decades.

7 High speed in a tank had other benefits too, such as the ability to catch up to slower tanks and thus play a role as a tank-destroyer.

It was with typical blind zeal that Christie attacked one of the main problems with tanks — getting them to and from the battle-field, wherever that battlefield might be. In the early 1930s, Christie developed a solution to the problem: he supplied a tank with wings. Just imagine the enemy's shock when they looked into the sky to see an honest to goodness tank flying at them. Christie envisioned a mechanized flying monster — a griffin with the body of a tank below and the wings of an airplane above.

He had a couple of ideas with respect to the flying tank. One was that the tracked machine would be outfitted with detachable wings and tail and would accelerate up to flying speed, at which point the propeller would get turned on and the additional thrust would get it airborne. To land, the pilot would get the tracks going to match the flying speed, then, when the tracks made land contact, he would brake as he would in a truck. Thus grounded, the wings and tail would be released "in a jiffy" by pulling a lever and dis-carded. The monster would then zip off with its "guns blazing" in ground assault.

Christie's ultimate idea of airborne deployment, however, was a portable tank with a lift-and-drop plane system. He envisioned a bomber with a ventral attachment system that would fly low, attach to a tank in mid flight, lift it, transport it to its new location, fly low, detach it, and then go and get another tank. Of his invention Christie said, "The flying tank is a machine to end war. Knowledge of its existence and possession will be a greater guarantee of peace than all the treaties that human ingenuity can concoct. A flock of flying tanks set loose on an enemy and any war is brought to an abrupt finish."

Remarkably enough, this idea caught the imagination of Lew Holt, a *Modern Mechanics and Inventions* journalist, in 1932, who seems to have spent a few happy hours with Christie, imagining the greater benefits to society of such a system. The two reasoned that Christie's invention could diminish the pesky problem urbanites have when traveling to the outskirts of their cities to get on a plane. With Christie's plan, air passengers could board a bus in the center of the city, be driven to the airport, get attached to a plane that picks them up — bus and all — fly to their destination, and repeat the process, in reverse, upon arrival. Imagine:

> Glancing through the windows which surround this compartment, we notice numbers of other plane-cars. Some are standing at their respective platforms, others are arriving or departing through the low doorless exits of the vast hall. At 5:15 to the dot our car glides smoothly forward and we pass out of the hall onto a wide high-speed road which leads directly to the flying field.
>
> Traveling at an even sixty, we arrive at the field to see a number of curious looking planes lined up before the service hangars. They appear to be all wings and tail and are lined up over sets of trough-like tracks which serve to guide the wheels of the cars into exact position as they arrive to be hooked up to the plane units. Into one of these sets of tracks our car wheels presently glide, the car stops and the ground crew busies itself locking the connections into place. A few moments pause and then the motors of the plane, which have been idling as we approach, burst into a mighty roar and in what seems to have been but a few minutes since we entered the car at the central depot we are sailing high over the city. . . .

In order to make his dreams of flying tanks a reality, Christie worked hard at making tanks as light as possible. His M1932 had a 750-horsepower V12 engine and twin duralumin wheels with pneumatic tires. Apparently, it could travel at speeds of 120 miles per hour (193 km/h) on wheels and 60 miles per hour (97 km/h) on tracks. The tank weighed 5 tons (4.5 t), and Christie boasted he could get it down to 4 tons (3.6 t). No one in the United States military showed an interest in the M1932, and the Soviets did not want it either, though it was likely Christie, not his tank, that they were avoiding.

As it turned out, weight was the least of the problems to be overcome in getting a tank to fly. We know this because, as wild as it might seem, there were some who believed in flying tanks then,

just as there are some cryptozoologists who believe in flying mon-
sters now. The Russians were particularly interested in the idea.
Based on Christie's 1932 design, they fitted biplane wings and a
twin tail to a T-60 light tank and flew it in a test flight. Called the
Antonov A40[8] after its designer, this tank was lifted into the air by
a tow plane and was released midair. It then glided into position,
landing on its tracks. The wings and other flying paraphernalia were
subsequently removed. A test flight was completed successfully in
1942. The pilot landed it and drove it back to base, quite enthusiastic about its performance. The experiment was discontinued however, perhaps because there was not a sufficiently powerful aircraft
to tow the glider up to the required speed when it was equipped
with men and guns. The experimental tank was flown completely
stripped of its all armaments and extraneous parts, including head-
lights and extra fuel. It was a shell with a pilot inside.

The Russians were not the only ones to get wind of the idea.
The Japanese also had an experimental flying tank, called *kuro*
(meaning "black") or *sora* (meaning "sky"). It doesn't appear as
though this chimera got far off the ground either. The British, too,
experimented somewhat with a glider that could carry a detachable
light tank. Called Baynes' Bat, the glider, or "carrier wings," was
tested as a half-scale model, but that is all. The British did transport
light tanks by air into combat in the Second World War, in the
hold of a large glider, the General Aircraft Hamilcar.

To some degree, we can only speculate why the flying tank pro-
totypes failed in testing, but to get some knowledgeable opinions,
I turned to the military enthusiasts on the Armchair General
Weapons and Warfare forum. Here are some of their responses:

Wedfactory: It was determined to be too hard on the crew inside the
tank, not to mention they now had to train tank drivers how to fly. (At
the time the tank crews were not known for being the best and bright-
est.) It was realized with the tank empty it was nearly deadly to the

---

8 Also called the KT40.

crew. If it had a full combat load it would have worked better as an asteroid than a gliding tank.

Mountain Man: Landing speed was fast enough to burn out the running gear on the roll-out. If the speed were a little too high, or the ground a little too bumpy — extremely common occurrences in combat ops — the lightly made running gear of the K-[6] couldn't handle it, according to some Soviet reports. The tank commanders would not, after all, have been very good "glider" pilots and no one would have prepped the landing zone beforehand.

Ace: [A] 5-ton tank would probably have to be made of cardboard.

DodDodger: The factor that killed the KT-40 was that the Soviets didn't have a towing aircraft powerful enough to overcome the tank's drag. On the single test flight, the tank had to be cut loose prematurely to prevent the towing TB-3 from crashing.

Paul Soderman, a retired NASA aeronautical engineer, has this to say about the feasibility of flying tanks:

A 5-ton flying tank is certainly possible. A Boeing 747-400 has a take-off weight of 875,000 lbs or 437 tons. Its wingspan is 211 ft. So a 5-ton aircraft would only need a fraction of the 747 wing. Any wing could be detachable, but the weight of the junction makes it undesirable unless one has a very, very good reason to take the wings off; I guess needing a tank would be a good reason. But think of all the problems related to putting a flight control system in a tank, with a tail, a propulsion system, etc. Sounds very impractical. Tanks have enough trouble surviving on the ground.

Robert McKinley, an ex-NASA engineer who is now director of engineering for a company that sells mobile drilling rigs, which he describes as "vehicles that are much closer to tanks than to aircraft," elaborates on one of the technical difficulties of a winged tank, transferring power from the tank's engine to a propeller during take-off:

> [I]n the 1920s–30s, the most likely avenue would have been a mechanical power transfer system composed of gearboxes and/or chain drive(s) and shaft(s) connected to the tank engine. The power transfer system would have to make at least two 90-degree turns assuming that the engine in the tank is transversely or longitudinally mounted as in modern cars: one turn to take the crank shaft motion and turn it vertical, then another turn to take it back from vertical motion to horizontal rotation for the prop. Each turn adds inefficiency to the drive system (say ~10 percent per turn as a reasonable guess) as well as significant weight. . . .
>
> Each turn via gear or chain is also an opportunity for mechanical failure. Each turn requires a mechanical connection such as Cardanic linkage (u-joints) or bolted flanges. Connections have to transfer torque (in this case), but they also must survive significant vibration and perhaps mechanical shock loads as well. Balancing the drive shafts would be extremely critical, particularly at the scale of this vehicle. The tank's engine must have a "power take off" of some type that allows connection to the prop drive but doesn't jeopardize the drive system on the tank. Reliability might have been a lower priority, since the wing/prop side of the system was somewhat disposable, at least in wartime, but it cannot be too low of a priority if you have a human crew. In addition, the output speed (RPM) of the tank engine must be tailored to the speed (RPM) required for the propeller. This would have to be built into the tank transmission and the gear/chain drive ratios. The connection at the tank would also have to be easily disconnected after landing (another opportunity for mechanical failure).

Generally speaking, a flying tank is difficult to accomplish because it is attempting to do two things at once. Tanks are ground vehicles and are designed as such. Aircraft, on the other hand, are designed to fly. McKinley explains the consequences of trying to be a tank and an aircraft at once:

[A]ircraft (be they tanks with wings or otherwise) must be addressed as a *system*. . . . When the power, fuel, and landing gear subsystems of the aircraft also happen to be a tank, systems integration between tank and aircraft becomes the determining factor for the success of the system as a whole. . . . Each part of this overall flying tank system has to be designed with the goal of flight in mind. For the tank itself, this means that it is probably over-powered and under-armored for its war-fighting mission. . . . The aircraft has to take along all the parts of the tank, even though much of it is counter-productive from a flight performance stand-point. For the integrated tank plus aircraft subsystem, the overall result will be an aircraft that is probably marginal in almost all respects (over-weight, under-powered, poor handling qualities, limited range, etc.).

The Russians did not entirely give up on the idea of airborne tanks. They turned to parachutes; their BMP and BMD tanks are referred to as airborne combat vehicles (ACV). The modern video footage available on YouTube of tanks being tossed out of an aircraft in midair like neat, tiny parcels is quite astounding. The United States and other countries have dropped tanks from planes at various times in various ways too. The soldiers on the Weapons and Warfare forum relate their experiences:

peter sym: When I served in armor in the mid-'90s we heard some horror stories about the Russians fitting BMPs with parachutes and dropping them out of aircraft with the crew strapped in the vehicles ready to fight. The BMPs always landed far too hard and the crews suffered terrible spinal injuries. Clearly the idea of getting armor onto the battlefield by air persisted for a long time in the Soviet Union.

JSMoss: Reminds me of the experiments of airdropping the M551 Sheridan (known affectionately to those of us who served with them as the broken electric war machine). During the Vietnam era they installed several gun emplacements at Fort Campbell by trying to airdrop the M551s, which embedded themselves up to their turret rings in the ground. Fortunately they were not so foolish as to put crews in them.

Scout32: The M551 Sheridan AR/AAV was used by the 82d Airborne Div from 1969 till 1997. The United States Army no longer has any air droppable Armor since the retirement of the M551 and the ending of the M8 Buford Armored Gun System, which was supposed to replace it.

Engineer Paul Soderman comments: "Parachuting a tank is difficult because of weight. No matter how good the parachute, the terminal velocity depends on weight. At some point the impact is just too destructive without elaborate energy absorptive systems like NASA's balloons used for the recent Mars landing. Tanks are tough, but they contain optics and electronics not to mention shells."

Today, if the military really needs to get a tank, or AMV (armored military vehicle), into a tight spot, they can transport a light one with a heavy lift helicopter, such as a Chinook, Stallion, or Hercules. A light tank can fit inside one of these highly specialized helicopters or can be carried in a sling below it. Otherwise, most tanks are transported by rail, if possible, or by ship, cargo plane, or HETS (heavy equipment transporter systems).[9] The United States stopped using the last air droppable tank, the M551, in 1997. It seems that getting tanks to the battlefield has always been a challenge. And one can see how Christie's simplistic solution might have made some sense at the outset, even though it was not feasible in the details. Two machines designed to do two very different jobs cannot be stuck together and expected to perform both jobs just as well as they do apart. The griffin is a myth.

Occasionally, what seems impossible really is impossible.

---

9 HETS is a tractor-truck and semitrailer. For more info and photos see www.fas.org/man/dod-101/sys/land/hets.htm.

MAKE SURE YOU AREN'T THE ONLY ONE WHO THINKS IT'S A GOOD IDEA

GEORGE DAVISON'S
POPCORN
VOLCANO

1963-

This concept illustration depicts the Volcano Glowing Microwave Popcorn Bowl on display on a retail store shelf.

Finally! Someone has done something about those annoying bolt caps on toilets — you know, the little plastic knobs that always pop open or fall off altogether? His name is George Davison.[1] He's an inventor working in Pittsburgh, Pennsylvania, with a team of more than 80 employees. Together, they have developed dozens of products now sold to consumers like you and me — like the toilet bolt caps that screw on and don't pop off, as well as other nifty gadgets: a purse rack, personal air fresheners, children's dinnerware, and an RV leveler. Davison Inc. develops about 200 new products every month, about 10 of which are successfully licensed.

Davison started his invention business in his grandfather's basement, at the tender age of 26. His company, Davison & Associates Corporation, turned a profit three years later, and he hasn't looked back since. Davison won his first Industrial Design Excellence Award, for an oil filter gripper, in 1996. His Swiss Army Whistle Knife followed, and Home Depot took on Davison's brand as a private label. The output of his company exploded when Davison made the transition from inventing and licensing his own products to building a team of inventors and developing a system, dubbed "Inventegration," to take the ideas of others and turn them into viable commercial items. Some inventions brought to market by Davison Inc. have won international recognition from the design industry, like the Hot/Cold Therapy Wrist Brace, the BikeBoard, the Jack 'N Stand, and the Hover Creeper, the latter having gleaned the top concept award from the Industrial Designer's Society of America.

Right out of *Back to the Future II*, this "hovercraft" allows a mechanic to float on a film of air underneath a vehicle, without uncooperative wheels to reduce maneuverability. To understand the benefits of such a device, imagine trying to maneuver a shopping cart laden with hundreds of pounds of groceries — while lying

---

1 George Davison is a controversial figure in some invention circles; his company assists others to develop and market their inventions, including the toilet bolt caps. The information here comes from a phone interview with Mr. Davison in October 2007.

GEORGE DAVISON

185

on your back with a wrench in your hand and a ton of steel above
your head. Then try to put your muscle into that wrench to undo
an old rusty nut while under that ton of steel, while on wheels. The
Hover Creeper allows mechanics to move around when they want,
where they want, and stay put when they want. It plugs into an air
compressor, an item already found in most garages. Forty pounds
per square inch (276 kPa) of pressure will keep a man weighing 300
pounds (136 kg) off the ground.

The biggest challenges to overcome in making a workable
Hover Creeper were twofold: develop an air film that could support highly variable weights and sizes, and do so without irritating
air noises. A series of regulators and bladders within the creeper to
distribute the airflow do the trick. Davison says a hover skateboard
is possible, but would be heavy! He's interested in working on more
pragmatic ways to utilize hover air film technology — a hovering
mechanic's chair is next.

It is not unusual for great inventors to turn their minds to
whole-scale improvements of the way things are done. Henry Ford's
manufacture assembly line and George Washington Carver's holistic bid to assist the small Southern farmer are examples. George
Davison, in a similar big-picture vein, has turned his mind to ways
of improving the commercial product invention process itself. He's
focused on what he calls "enabling technologies," which assist
human beings to tap into whatever fuels their creativity. The result
is a 60,000 square foot (5,600 m²) facility called Inventionland.
Part reality, part virtual reality, Inventionland is the workspace of
his inventors, and it is unique in the world. It is a creative environment that has been carefully planned to tackle what Davison
describes as "the human challenge" — the ebbs and flows of the
human mind. His solution is to envelope employees with positive,
uplifting vibes and surroundings that tickle the core of their inspiration. Designers working in the automotive industry sit inside a
model Indy 500 track. The infant department works in a giant crib
complete with giant mobile. The outdoors department has a rock
cave. The toy department meets on a pirate ship. Launched in

2006, Inventionland is a surreal office composed of high-tech computers in a Disneylandlike setting, the antithesis of the cubicles in *Office Space*.

Does it accomplish its goal? In a 2006 interview with the *Pittsburgh Post-Gazette*, "Inventionman" Matt McClatchey, whose job entails dreaming up games, says so: "your mind has complete freedom to do whatever you want to do here." It is too soon to tell how this translates into quantity and quality of invention. Whatever you think of it, it cannot be dismissed as the mere playing out of childhood fantasy; the concept of welcoming, positive architecture designed to enhance creativity is echoed in one of the premier theoretical research institutes in the world, the Perimeter Institute in Waterloo, Ontario. There, world-class academics bask by wood-burning fires amid espresso machines and roomy sofas with blackboards at the ready should inspiration strike. We can only hope that the culture of creating such warm and fuzzy employment spaces trickles down to the rest of us.

Davison shares many of the personality traits common among the greats. His penchant for tinkering with machines, his compulsion to understand how things work, and his head for business were present at an early age. In elementary school, George and a pal had a mini-manufacturing plant going in their garage to sell "Moke" — a mixture of coke and milk — to neighborhood kids. At about age 12 or so, he had a candy business operating out of his school locker. Each day he bought lollipops and other assorted confections at the store in town and then sold them at a handsome profit to rural kids at school who didn't have the same access. Foreshadowing things to come, teachers blocked his access to consumers when they started discovering globs of chewing gum stuck under desks.

Like most inventors, Davison has had his share of failures. His invention career started with a failure, when he sank all of his energy and resources into a toothbrush sanitizer, only to watch another company put a similar product on the shelves before he could. Failure in the lab is one thing; Davison and Edison agree this is a positive, motivational opportunity to try again. Davison says, "I see a failure as a mirror. For me, in inventing, this is especially important. You've taken an idea and built it into a well-constructed, fingerprint-free reflection of yourself. Then it fails. It smashes into hundreds of shards of mirror. To some it's a mess, a disappointment. To me, I see hundreds of new ideas staring back at me. I just choose a new path and go on from there."

However, failure for a product to sell or be accepted by the public is another matter. His Tree House Swing Set only failed because the bank funding the manufacturing company pulled the plug. The Screaming Bug Zapper bombed when the buyers at the major retail chains refused to see its potential, despite its 10 anthropomorphic Looney Tunes–esque death songs. But Davison is still scratching his head over this one — his Popcorn Volcano. The public should have loved it. It was a Walmart natural; it should have sold. It didn't. All because of what Davison calls the "human factor."

The Volcano Glowing Microwave Popcorn Bowl promised its owner "fun watching volcano erupt for 3 to 5 minutes." As the corn inside the volcano heated up, the volcano glowed red, until the heat caused the kernels to explode out of their shells. The cooked, fluffy insides expanded in volume, and the whole thing erupted out the top of the volcano's crater. The stuff of family movie nights? The buyers at American's retail giants didn't think so.

The volcano's glow came from a small tube filled with neon. Microwaves excited the gas causing it to glow brightly, and immediately. The idea came from the use of florescent bulbs in testing for leaks in microwaves. The volcano's neon lightbulb was encased in transparent silicone to make it safe to touch. The microwave heated a stainless-steel plate at the base of the volcano, which caused the corn to pop. So far, so good. So what made this invention a total flop? Davison's "human factor" in this case was the buyers and decision makers from the retail giants, which largely control what shoppers see on the shelves. Davison thinks that the "gatekeepers," as he calls them, may not always be the best people to judge whether an idea is a good one or not — or even whether the public would buy it.

It is a logical argument. All kinds of personality traits might attract a person to the job of buyer at a retail store, and the capacity to envision change may not be one of them. Davison points out that a noncreative type is less likely to back a new idea. Indeed, the criteria listed in job ads on Monster.com by those hiring buyers emphasize budgeting, negotiation, and people skills. Words like creative, visionary, and innovative are nowhere in sight. From a buyer's perspective, any new product is vying for shelf space with other products, perhaps with an already proven track record. New items are not necessarily better. Buyers also have agendas and strategies put upon them by their superiors; they are not totally free to follow their own instincts. Budgets are a consideration as well; a great invention may not bring a manufacturer or retailer as much profit. Even small differences in the bottom line could prevent the public from ever having access to new ideas, no matter how much they might love them.

And, of course, wherever there is humanity there is human error. A buyer might be in a bad mood and not emotionally receptive to anything fun and frivolous, like a volcano popcorn popper. Or, they may simply misjudge the public's tastes. There are plenty of examples of this happening, such as the nine editors who rejected *Harry Potter*. What if J.K. Rowling had stopped trying after the first half-dozen rejections? A worse thought: what masterpieces have been lost to us because their creator gave up trying to convince the gatekeepers?

The capacity for gatekeepers to determine what the public can buy might be changing, thanks to the Internet. As never before, consumers on a mass scale have direct access to producers. The direct link has already transformed the music industry, along with the movie and print media industries. Chris Anderson, editor-in-chief of *WIRED* said, "the mass market is yielding to a million mini-markets." It is now possible for innovators to bypass the gatekeepers and get their message — and their product — to potentially millions of people via YouTube and other sites. The revolution has just begun.

Determination and a strong belief in one's work is definitely a trait that characterizes great inventors. And when an inventor is right and the gatekeepers are wrong, determination has historically been very important to eventual triumph. In the 1930s, Clarence Darrow invented a board game. People liked it, and he made copies of it for friends and neighbors. When he'd made a hundred copies he approached Parker Brothers, who told him all the reasons that it wasn't any good — it took too long to play, and it was too complicated. There were 52 "design errors" identified by Parker Brothers in all. But Clarence Darrow thought these experts were wrong. So he continued to make copies of the game himself and sold them, one by one, store by store. Two years later, Parker Brothers changed their minds, and Monopoly has been with us ever since.

With prolific inventors, determination is an obvious factor in their achievements. Being prolific, however, is also conducive to having plenty of duds. When ideas grow on trees, it is easy to drop a wormy apple in favor of other, yet unblemished fruit. In modern times, when people like George Davison have 80-plus inventors working for them and invention-development companies like his receive up to 55,000 ideas from the public each and every month, it is easier than ever to give up on an idea in order to pursue greener pastures. The scrap heaps of acclaimed inventors are a lucrative place to look for hereunto unused world-changing ideas.

The luxury of not needing to continue to champion their castoffs is one reason why great inventors may have great flops. But there are other internal weaknesses at work. Creativity is an interesting mental process; some say it is *the* defining property of the human mind. As such, psychologists have studied creativity with an eye to figuring out how we work. As it turns out, creativity is associated with a slew of cognitive and personality traits, including schizophrenic personality, rebelliousness, difficultness, androgynous character, sexual libido, and arrogance. Furthermore, creative thinking is something altogether different than rational thinking; it is an unrelated process in the brain. Hence, the propensity of creative thought to present itself when we are half-asleep or have our "rational switch" turned off.

Davison acknowledges that when he's in an inventive mental mode, he is not open to communication with others. In a sense, he cannot hear the voice of outsiders because his focus is so strong. If, by nature, successful inventors are generally out on a mental limb, operate outside logic and are also by nature stubborn, rebellious, and proud, we need not look any further than an inventor's own mind to explain how greatness can coincide with not-so-great ideas. Consider the following scenario: while thinking outside the box, an inventor has a bad idea. Because the creative personality is open to ideas that others are not, even ideas that are illogical, the inventor does not perceive the idea realistically. And he or she certainly does not perceive the idea as others — such as potential customers — do. And then, the inventor stubbornly and determinedly sticks to this illogical, undesirable idea, simply because it is in his or her psychology to do so.

Like love, creativity is blind.

PLEASE THE BUYERS
and THE SELLERS

JEROME LEMELSON'S
FLYING BALLOON
1922–1997

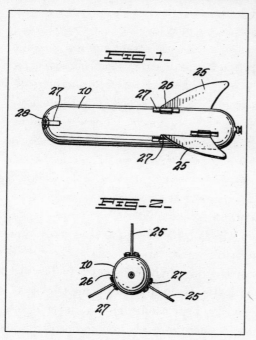

One of Lemelson's 605 patents was for a flying balloon, officially an "Inflated Aerial Toy." It was granted in 1956.

The first things that great inventors play with are toys — just like all other children. Buckminster Fuller and Jerome Lemelson both spent a great deal of time playing with model airplanes as boys, Jerome flying his in Staten Island, New York, where he grew up. However, while the mature Fuller focused his innovation efforts largely on basic human needs like housing and transportation with a view to saving the world, Jerome Lemelson was more of a free spirit. Jerry, as all those close to him called him, was the kind of inventor who gave his imagination free rein, all day, every day. And what an imagination it was!

After graduating from New York University with two Masters degrees, one in aeronautical engineering, the other in industrial engineering, Lemelson spent his days working for Republic Aviation. His brother Howard, who shared his one-bedroom apart- ment, shed some light on what Lemelson did at night: almost hourly, night after night, Lemelson turned on the light to scribble down some new revelation. Howard Lemelson recalled how each morning he was shown the results of the night's ruminations — half a dozen or so new ideas for inventions, written out in a notebook, which Howard was asked to sign as witness to their origin. The habit of writing all these ideas down never left Lemelson; he carried a note- book with him always, just as Thomas Edison had done 50 years before him, and Leonardo da Vinci had done 450 years before that. Lemelson is remembered sitting under an umbrella with the note- book in tow on family beach outings, in between jaunts in the water.

The breadth of patents that emerged from Jerome Lemelson's notes is astounding. He averaged one patent application per month for a period of more than 40 years. One early, important patent was prepared and submitted as a 150-page document to the patent office in 1954. It described machine vision, which is a camera on a machine coupled with a digital image to which the machine can compare the live object. This simple concept is used heavily in quality control in modern manufacturing processes, though digital technology makes it possible to now compare 3-D images rather than the 2-D imagery that existed when Lemelson first dreamed it up. Another related Lemelson concept is bar code reading, in which goods are given a symbol comprised of bars and spaces that represent numerical data that can be automatically "read" by a computer. There are many other manufacturing process inventions in Lemelson's patent list, including his most recent patent at the time of writing, granted posthumously in March 2008: "a machine tool apparatus is provided employing a plurality of tool heads which may be sequentially or simultaneously automatically controlled to perform preprogrammed operations on either the same or different workpieces."[1]

Lemelson's audiocassette drive innovation made the Sony Walkman possible; he also thought of the camcorder and the videophone. He made contributions to drug administration, reflective highway markers, car warning systems, fax machines, ATMs, friend-or-foe systems for the military, and audio/video devices embedded in a photograph. As an individual, Lemelson made a remarkable contribution to modern technology through his long list of patents. There are 605 of them (and counting), making him the most prolific independent inventor of the modern era.

Lemelson's illustrious career as an independent inventor in this day and age is a reflection of a person with tremendous strength. Modern times are not conducive to the success of the independent inventor, which is reflected in the dramatic and consistent drop over the decades in the number of patents awarded to independent inventors compared to corporations. In the early years, Lemelson worked as an engineer for a smelting plant in New Jersey by day and worked on his inventions after hours. When he quit to pursue his inventions full-time and tried to license them to make a living, his wife, Dorothy, supported the family. She continued to do so well into the 1960s. Lemelson's inventions took a very long time to pay off, despite their close connection to big consumer dollars. It wasn't until the 1980s, more than 30 years into his career and with more than 360 patent files under his belt, that Lemelson's financial

---

1 US Patent Number 7,343,660.

rewards began to reflect his contributions. The struggle of the early years never ceased, as Lemelson took on the system, head on, earning him a reputation as "the lone wolf."

One example of these early struggles took place in Lemelson's late twenties, when he was busy inventing and doing all the work required to file a patent, by himself, at his own expense. He had an idea for a cardboard mask that could be made from the back of a cereal box. He wrote to cereal giant Kellogg's to see if they might want it. They refused. A couple of years later, while grocery shopping, Lemelson saw his idea on the back of a Kellogg's cereal box. He was flabbergasted and, with typical Lemelson gumption, sued them; Kellogg's response was that they had printed masks on the backs of their cereal boxes numerous times before (seven to be exact), as far back as 1947, and held patents for these masks. Lemelson lost the case. This would not be the last time that Lemelson would find himself in a similar conflict with a big company. Word in the invention biz is that corporations generally have something against NIH (not invented here) products. Many companies don't like independent inventors, don't like paying licensing fees, and don't like giving financial credit for others' ideas if they can make the idea their own.

Longtime friend Jack Gilstein was there with Lemelson through the hard times. He explained, "It would strap him financially just to build a model of an idea, but most big companies weren't interested in dealing with independent inventors. Their attitude toward Lemelson was: 'Go away, and don't darken our door again.'" Lemelson's son Robert was reminded recently of the impact this had on his father and family when he watched *Flash of Genius*, a 2008 movie that tells the story of Dr. Robert Kearns, who invented the intermittent windshield wiper then fought for three decades with the Ford Motor Company who claimed it as their own. The depiction of the frustrated inventor was all too familiar to Robert Lemelson. He sat in the theater for 15 minutes after the credits had rolled, too emotional to move.

Lemelson was never one to accept injustice unless he had to. In addition to the expenses of invention development and patent filing, he added a hefty legal component to his budget. He sued Mattel for the flexible track toy that was his idea.[2] He also took on the big four automakers in 1989, for the use of his bar code reading technology; the case was settled out of court. Machine vision and bar code reading technology is an area in which Lemelson has 14 patents. To protect them, he and his lawyer, Gerald Hosier, approached manufacturing companies that use this technology in their products and asked for a small royalty. Industry is so dependent on this technology and it is so widespread that this amounted to literally thousands of companies. About a thousand of them responded positively to the request — to the tune of about $1 billion dollars. There are still more court cases pending. As Gerald Hosier has commented, "What Jerry is doing is standing up for the civil rights of inventors."[3]

Jerry Lemelson never quit. When liver cancer struck in 1996, he continued his battle. He was not finished inventing, and he even stepped up the pace during his illness, filing an astounding 40 patents in the last year of his life, many of them pertaining to medicine. Cancer cut Lemelson's imagination short; he died in 1997 at the age of 74. But he had already found a way to ensure that his battle for independent inventors and the contributions that society receives from their imaginations never ceases. He and his family instituted a foundation in the 1990s that "support[s] inventors and entrepreneurs to strengthen social and economic life." Through MIT, a $500,000 prize is awarded each year to an independent

---

2 After 22 years, Lemelson won his case in 1989, but the decision was overturned three years later on appeal.

3 The opposing viewpoint is that Lemelson purposefully used the patent system to protect "submarine patents." He kept his patents in the review process by repeatedly filing small amendments, keeping his claim valid while he waited until the technology developed by others in a natural progression of thought and engineering finally resulted in the concept being widely feasible. Then Lemelson's patent emerged like a submarine to be granted, allowing him to reap the reward for the work of others.

inventor. As well, a hands-on education and research center was established at the Smithsonian National Museum of American History to highlight the contributions of America's inventors and foster innovation skills in youth. Other experiential education programs for innovation are supported at Hampshire College in Massachusetts and the University of Nevada, as well as smaller scale projects of a similar nature at hundreds of other colleges across the United States. There is an additional component to support innovation and invention for the developing world, with several programs that target "design for the other 90 percent."

While Lemelson was quite a warrior, he was also quite an elf. Biographer Robert Siegel describes his attic workspace in the family home in Metuchen, New Jersey, as "almost like Santa's workshop, filled with toys, some of which he bought to study, but mostly those he created." Lemelson surrounded himself with toys and loved to watch his sons, extended family, and other children play with them. He also liked to play a great deal himself. Lemelson's son Robert said, "He took the idea of play and transformed it into a lifelong vocation." Lemelson loved to ski, both downhill and on water, and did so his entire life. His official biographical sketch on the Lemelson Center website describes him as having "a sweet, gentle, playful side that was particularly evident when he was with children."

Lemelson's playful side was evident in a variety of toy patents. In fact, his first patented invention was a toy — a beanie cap with a propeller on top and a straw to blow hot air at the propeller. The year was 1953. Lemelson followed with 23 more toy-related patents in the 1950s, nine in the 1960s, and 25 in the 1970s. Over his lifetime, he had 60 toy-related patents, about 10 percent of his total output. Among the toy patents are some well-known items: bendable plastic car track (the same type made by Mattel Corporation for use with HotWheels) and a ball and target game that uses Velcro. There are also building units and powered vehicles, crying dolls, and ricochet noise-making guns.

Inventing toys was fun, but it was also pragmatic. The toy industry is gigantic; Mattel, the world's largest toy manufacturer, reported revenues of close to $6,000,000,000 in 2007. That's 6 billion dollars — and a lot of zeros. Hasbro was not far behind with $4 billion, and that figure is growing. Every year between five and six thousand new toys are launched in the United States. Lemelson purposefully targeted the toy industry in his earliest inventing days because he felt it was the easiest market in which he could get a start. It was part of his master plan: "In the beginning, I wanted to manufacture certain ideas I had in the toy and hobby field and become financially independent. After that, I planned to get my own lab and machine shop and develop my ideas further." Lemelson was not alone in deciding that the toy industry would be the easiest place to find success, and many independent inventors target this sector to this day.

Lemelson's elfin side may have had something to do with his invention prolificacy. Play is extremely important for creativity — and hence innovation. Play grows the physical capacity of the brain and develops the connections between motor functions, senses, and cognition. It provides practice in symbolism, abstraction, and learning. Not surprisingly then, children who are more playful are more innovative as adults.

Play in adults also develops creativity and divergent thinking, as well as enhancing relationships and arousal and improving mood, learning, and mental function. Some design firms know this and use it. The company IDEO, for example, encourages play as a regular course of their design work. CEO Tim Brown gave a talk about it for a TED Conference in 2008, describing the hows and whys of it. Innovation consultant Deanna Berg believes in the power of play in the workplace, and she offers helpful suggestions for working playtime into the office routine, such as organized activities like walking backward, bringing out the toy box for a one-minute toy break during meetings, or using costumes in training exercises. The result of this playtime is reportedly enhanced creativity — and greater job satisfaction. Psychiatrist Stuart Brown has taken the science of play further. Not only has he researched play and concluded it is critical to human health,[4] but he has also founded the National Institute for Play, part of whose vision is "a future in which school systems have used the knowledge of play to topple the current morass in K-12 education and a future in which public and private sector leaders have used play practices to reform organizational policies and create organizations capable of producing innovative products and services."

While Lemelson's inventiveness enriched the world of toys, his consistent occupation with toys may have also contributed to his more serious inventions. Senior Historian Joyce Bedi of the Smithsonian's Lemelson Center has pointed out the parallels between the toys Lemelson invented and the more technical inventions he was developing at the same time. For example, during the same period he was contemplating magnetic recording and reproduction devices, he devised a fishing game, a car-and-track toy, and a toy mine detector — all of which use magnetism. Similarly, applications of the properties of light for reflective thread and light-sensitive contact lenses were conceived during the same period that Lemelson dreamed up methods for producing optical effects in inflatable toys.

Among Lemelson's toy inventions are some inflatable playthings, including a rocking horse, a hobbyhorse on wheels, and a space capsule. There is also an inflatable target game with a full-size blow-up doll made to look like a sports figure. The doll has a ball-shaped cavity in its midriff that "catches" a ball if it is thrown just right. The flying balloon, officially called "Inflated Aerial Toy" on the patent application, has US Patent Number 2,763,958.

Lemelson's inspiration for his inflatable inventions may have come from his knowledge of aeronautics. Among his jobs as a young man, Lemelson worked as an engineer for an aircraft manufacturer

---

4 Brown's first study of the importance of play noted its absence among homicidal young males, including mass murderer Charles Whitman. Stuart Brown discusses the importance of play in a TED talk available online at: www.ted.com/talks/stuart_brown_says_play_is_more_than_fun_it_s_vital.html.

and also for a weather balloon company. He stated his motivation for the flying balloon invention in the patent; he aimed to provide children with "a flying toy which is simple in structure, easy to package, comparatively cheap and exceptionally safe in use." The balloon airplane is a kit, consisting of a long balloon, a set of adhesives, a weight to add to the nose of the balloon, and three fins with tabs that are stuck onto the balloon near the tail. With just the right amount of weight to balance the nose and tail and the fins attached symmetrically to guide flight, Lemelson's flying balloon can sail more than 50 feet (15 m) through the air when launched by a firm hand. His son Robert remembers trying them out. They flew splendidly, and he recalls the experience fondly. Robert says the airplane went along with another idea for an inexpensive toy made from balloons, the "balloon head." This kit consisted of stickers that turned the balloon into a wacky caricature.

Inflatable toys are a great concept and there are plenty of them on the market today, including several that reflect the strength of Lemelson's blow-up ideas. A rocking horse was developed by Ledraplastic in 1984, exactly 20 years after Lemelson received a patent for his idea. It retails for $40. Rawlings has a line of "Inflatable Target Sets" that are Lemelson's blow-up target patent come to life. There is a Touchdown Hero for football, Strikeout Kid for baseball, and soccer and basketball models too. All are widely available at Walmart, Target, and sporting goods stores for a retail price of $9. The flying balloon, however, only exists as a patent drawing.

Robert Lemelson says his father "did not believe in failure," but this was one invention that did indeed fail. No one bought the idea then, and no one has since. As far as the balloon flies, and as much as children might find it enticing, no interested parent can purchase this toy today. The closest a consumer can come is access to video instructions for balloon twisting. It appears making a balloon airplane out of those long sausage-shaped balloons is standard repertoire for birthday clowns, but the end result won't fly anywhere.

There are several reasons why Lemelson's flying balloon did not

succeed in bringing him any income, none of which has anything
to do with the frivolity of balloons and stickers. For one, the majority of new toys are invented by professional toy designers employed
full-time by toy manufacturers. There are more than 1,000 toy manufacture companies in the United States. This industry, generally
speaking, has the same "not invented here" problems as elsewhere.
For another, the flying balloon lacks some key ingredients in a successful toy; while it might meet the needs of the consumer, it doesn't
meet the needs of the manufacturer. In the Toy Industry Association's *Toy Inventor and Designer Guide*, the two criteria listed for
determining whether a toy concept is sellable are:

1. The toy can be manufactured for a cost that can "ensure a
    profit"
2. The production costs of the toy and its market-driven price
    point will bring an acceptable profit margin

Lemelson's flying balloon failed because it did not appeal to toy
companies. A company's goal is to make money, and this depends
on much more than the merit of the product. The flying balloon
may have been appealing, but manufacturers could not charge
enough to make a profit big enough to make it worth their while.
Though a perfectly good toy, there is no shortage of other good toys,
and children have not exactly suffered from its absence on the toy
shelves.

In other arenas, such an immediate–profit driven approach
to innovation has consequences well beyond what toys are brought
to market. Recent research into the nature of innovation in the pharmaceutical industry provides a case study. To explain the connection
between play and innovation in industry, let's first define "play." Over
the last decade, researchers have become increasingly convinced that
the process of innovation and invention *is* play. Play is defined by
the presence of a number of elements, including symbolism (repre-
senting reality with "what if"), meaningfulness (it connects experiences), action, pleasure, voluntary participation, rules — be they

implicit or explicit, and episodes characterized by shifting goals. Add to that the practice of collective skills and the element of chance. Many of these elements are present in innovation. A study published in 2008 examined the process of new drug development in a multinational company within the framework of play. Not only did the findings suggest that scientific drug development is a form of play — with the exhibition of individual and collective skills, rules but unknown outcomes, and the element of chance — but also that profit-enhancing measures are having a potentially negative effect on innovation. The study pointed out that the practice within the pharmaceutical industry of automating the development of new drugs as much as possible has detracted from the interaction between scientists and disrupted their *play*, with deleterious consequences for innovation. Today, as in other eras, society is struggling with how to reward its "idea men and women," while utilizing mass manufacture to bring goods to as many as possible.

The bottom line? Imagination and profit do not always play well together.

STANLEY MASON'S
CHINESE TALLOW
—TREE—
PLANTATION

• • • • • • • • • • • •

1921–2005

Fig. 6.
(PRIOR ART)

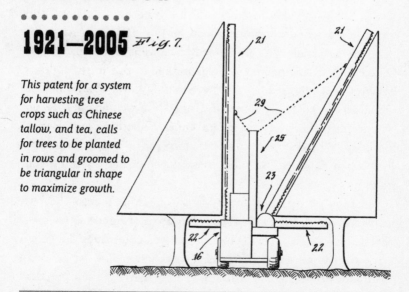

Fig. 7.

This patent for a system for harvesting tree crops such as Chinese tallow, and tea, calls for trees to be planted in rows and groomed to be triangular in shape to maximize growth.

It was always the ordinary, everyday American problems that intrigued Stanley Mason most. Like the day in 1949 when his wife asked him to change their infant's diaper. He looked at the square, flat disposable diaper and then at the round bottom before him, and decided there was a better way to do things. So he invented the contoured disposable diaper. Mason, in the five-decade invention career that followed, brought us many other commonplace household items. To name a few: the squeezable ketchup bottle, disposable surgical masks, granola bars, heated pizza boxes, dental floss dispensers, Masonware microwave ceramic cooking dishes, the "stringless" Bandaid packaging, the plastic underwire bra, and the contoured sanitary napkin.

Mason has an all-American persona to match his all-American inventions. He developed his first successful invention at the age of seven, when as a young boy in Trenton, New Jersey, his father refused to give him money to buy a fishing lure. Instead, his mother gave him a wooden clothes peg and suggested he fashion his own. He did. The wooden minnow was followed by numerous other matching wooden minnows, which sold for 26 cents each to his neighborhood pals. The grown-up Mason went to university, graduated with an engineering degree, and did a stint in the military, serving as a fighter pilot in the Second World War. Then he obtained several decent jobs with United States companies, designing products. He was fired several times, something he was quite proud of; as he bragged to a journalist in 1998, "You should always be fired in America." In total, Mason developed over 100 inventions and patented 60 of them.

In 1973, with the advent of Masonware, Mason began his own company, Simco Inc., which specialized in creating cosmetics, medical devices, and packaging for Fortune 500 companies. The company's offices were attached to his house, and he later wrote for an online newsletter about working at home, giving such practical advice as: set up personal boundaries, assign yourself regular work hours, dress for "the office," and make sure you have dedicated space with a door that closes and a sign to tell the household you are "at work." He freely gave inventors advice, too, dispensing gems like "The trick is to find something that the customer will want before you set pencil to paper!"

Stanley Mason described himself as "an inventor of ordinary, everyday products — not high-tech, but common useful things." He was certainly a practical man, so it was not *entirely* out of character when he turned his attentions away from the small and everyday to tackle one of modern America's biggest challenges — the need for energy resources. The scope of the problem may have been new to Mason, but he applied his same basic philosophy. He thought about the problem from first principles and developed what he considered to be a simple, practical solution. Mason believed the answer was the Chinese tallow tree, an introduction from Asia and common throughout several Southern states. He thought the tree would make an excellent plantation crop, with its main product being oil. Mason got the idea from his knowledge of the Second World War. He knew that Chinese and Allied forces had used oil directly from the tree as diesel fuel during the hard times of the conflict. He cited some of its earlier champions, including Benjamin Franklin, who sent some seeds in 1772 to Dr. Noble Wimberly Jones in Georgia with the note, "tis a most useful plant."[1]

It is true that the Chinese tallow tree (*Triadica sebifera*[2]) has a large potential yield of biofuel. Its seeds are 40 percent or more oil, and it yields about 500 gallons of oil per acre (4,733 L/ha) from its seeds. By comparison, an acre of soybeans can yield about 30 gallons (285 L/ha).[3] Soy is the most common biodiesel crop; the tallow tree is thus 15 times as "oily" as its biggest "competitor." Mason described the tree to be "like the pig. You use everything." The wax products that give the tallow tree its name can be used for making

---

1 Benjamin Franklin is credited with introducing the plant to the United States.

2 Also *Sapium sebiferum*.

3 Figures refer to Texas crops, as quoted by David Shermock, AgriBioFuels, Inc., from *Texas Agriculture*, (September 1, 2006), www.txfb.org/texasAgriculture/2006/090106/090106biodieselcorrection.htm.

candles and soap. The outside part of the seed can be used as a sub-
stitute for edible fats. Paints, varnishes, plastics, wood chips, and
cattle feed are among its other potential commercial products. The
wood is white and close-grained and can be used to make furniture.
The flowers are favored by honeybees and produce a pleasant, light-
colored honey. A black dye can be derived from the leaves. And, in
addition to its oil, the tallow tree grows quickly so it is also a source
of biomass to burn for further energy production.

Mason put his back into the Chinese tallow tree plantation
plan. He courted politicians, agricultural bureaucrats, and corporations. He got to work on the mechanics of growing, harvesting, and
producing, and he filed a patent for a system that improved on current methods of tree crop harvesting. He started a promotional
campaign to garner support. In 1981, he gave a presentation to a
congressional subcommittee — George Washington Carver style
— with a suitcase of Chinese tallow seeds as a prop. It failed to
really impress. He knocked on the doors of numerous corporations,
seeking business partners in the venture. In the decades before his
death, it had become his passion. He stated simply, "This is a project which must be done."

Mason did make some progress. In an article written for
America's Inventor, he said that his company's work "received inter-
national attention and significant direct and in-kind support from
the National Science Foundation, United States Department of
Agriculture (USDA), Department of Energy (Energy-Related
Inventions Program), the State of Hawaii, numerous academic
research institutions, multinational interests such as Unilever,
Anderson Clayton, Deutz Diesel, Johnson's Wax, and agricultural
plantations in Hawaii." In 1985, the USDA Office of Critical
Materials put the Chinese tallow tree on a list of potential commercial species. In the late '90s, Mason reported that his company
was testing prototypes of the harvester he'd developed on the
Hawaiian Islands.

Mason's patent for a row-crop growing and harvesting system[4] was an integral part of his plan. The goal was to maximize the growth of the tallow tree crop while developing an efficient, mechanical method of harvesting to replace the traditional hand-picking done in the tree's native lands. The system he devised required the trees (a tea tree plantation is the example given in the patent) to be grown in rows with just enough room in between to ensure the shadow of one row does not fall on the row next to it. The tree was to be grown and cut so that it took the shape of a right-sided triangle, with the hypotenuse facing south. Mason further explained how a tractor implement could be designed to cut and harvest the triangle trees. The implement would consist of long cutting blades — on one side a fully vertical blade to cut the "backs" of the trees in one row, and on the other side an angled blade to cut the slanted, south-facing sides of the trees in the adjacent row. Additional short blades on the implement would extend lower down on both sides to cut the bottoms of the trees.

When asked what they thought of this device, a John Deere PR representative responded with "farmers don't want trees in their fields." Tea is still picked by hand throughout the world, so the idea has not caught on in that sphere. In theory, the concept is sound, but engineer and tractor enthusiast Jeff Miller is "dubious that it would work practically. Mainly because plants tend to grow where the light is and would not keep the nice shape proposed." As Miller points out, however, tests would need to be conducted before anyone could make conclusions; those tests, as far as I'm aware, have not been done. Dr. John Cline, a professor at the University of Guelph specializing in pomology, the science of growing fruit, has said the crop arrangement is not only sound but has been put into practice in orchards, though the rows need to run north-south

---

4 US Patent Number 4,327,521.

and not east-west, as Mason proposed. This simple mistake high-lights Mason's inexperience in agriculture — as his arrangement has half the exposed row crop facing north, receiving little sunlight.

Harvesting methods aside, the Chinese tallow tree shows some promise as a biofuel crop, at least theoretically. However, there is a potential problem. This tree is an invasive species, which outcompetes native species of deciduous trees. It spreads; it overtakes; it destroys. Since the 1700s, this woody plant has subsisted on United States soil, but for two hundred years its spread was not a threat. The United States Department of Agriculture, unaware of its potential to do damage, helped it become established and increased its range. In the early 1900s, they planted it along the Gulf Coast with the best of intentions, considering it a possible oil-producing crop. The tree now grows in most counties in Texas, Florida, Louisiana, and along the Carolina coast. It is also established in California and Arkansas. It occurs commonly enough that colorful names for it have sprung up regionally, like the popcorn tree, chicken tree, and Florida aspen.

The Chinese tallow tree is one of the "dirty dozen."[5] Today, it is on the Nature Conservancy's list of most wanted invasive species, and Florida, Louisiana, Mississippi, and Texas have formal plans to eradicate it in sensitive areas and manage it in order to limit the damage. And the damage can be extensive. The Chinese tallow tree has destroyed large tracts of prairie in Texas and marshes in Louisiana, and it outcompetes native tree species in some forest habitats as well, wiping out other plants and animals in the process. It thrives in a wide variety of habitats in both sun and shade: grass-lands, swamps, brackish waters, and upland forests. Field tests and models have shown that the tree is capable of spreading and thriving well beyond its current range to areas as far north as Illinois, New Jersey, Maryland, as well as patches along the West Coast. It is known to change soil chemistry, altering the habitat in the long

---

5 A "dirty dozen" list is a common way to define the most important invasive species.

term and the plants that can grow there. It has displaced native deciduous plants in many areas, where it grows in monocultural stands in vacant lots, abandoned fields, and in hedgerows. Many animals, including most (but not all) insects that rely on native plants for food, are displaced by it, although some songbirds use it. It is considered by the USDA to be responsible for the near extinction of several grassland bird species. The sap in its leaves and berries is poisonous to humans, and even touching it can cause skin rashes in some people. In short, the Chinese tallow tree is a nemesis of biodiversity. Like many invasive species, the Chinese tallow tree has an impressive reproductive rate. It starts to reproduce in three years or less, produces an average of 100,000 seeds a year, and continues to do so for about a hundred years. Stumps re-sprout, and roots send up new shoots. In other words, once established it is "virtually impossible to eliminate."

There are a couple of tiny silver linings to the tallow tree's invasion. For one, the species has proven useful as a study model for learning about the nature of invasive species in general. This is pretty important stuff, as invasive species are one of the prime forces evoking change in ecosystems globally. While trading plants between regions of the world has been occurring for millennia, the explosion of global travel and trade, as well as other hefty environmental pressures, has brought invasive species to the forefront of environmental issues. The Chinese tallow tree has been the object of a recent study to help determine exactly what it is about invasive species that makes them so good at outcompeting the natives. Is it that they have fewer natural enemies? Or, are they better able to recover from the damage done by those enemies? Is there something in the genes of the invasive individuals that makes them so good at taking over?

To test these ideas, Jianwen Zou grew Chinese tallow tree saplings in pairs in the United States out in the open, where animals and pests could eat them. Some sapling pairs were from United States trees, and others were saplings that came from modern-day Chinese tallow trees in China. The research compared the genetics from China to the stock that has successfully become invasive since its introduction 300 years ago. The results were intriguing. Both the American and Chinese saplings had equal amounts of pest and disease damage, and the American saplings had greater amounts of feeding damage from herbivores. Animals seemed to prefer them. However, the American saplings responded to the feeding with a speed of recovery and growth not seen in the Chinese saplings. The invasive Chinese tallow tree in the United States is thus different from its Chinese ancestor in a critical aspect that allows it to spread rapidly: it has a superior ability to tolerate damage and grow quickly in recovery.

Sound like a likely candidate to solve some of America's problems? Mason had plans other than eradication for the Chinese tallow tree. His vision was to grow Chinese tallow tree plantations in Hawaii as a replacement crop for the suffering pineapple plantations there. He wanted to kill several birds with one stone, providing economic diversification for Hawaii, as well as a source of energy for a state that does not have any energy resources of its own.

There is some continuing interest in the potential of Chinese tallow for biofuel. David Shermock of AgriBioFuels Inc. in Houston, Texas, thinks it could be grown in orchard situations on scrubland not currently used for any crops to an extent that would make a difference both to the United States' "fuel security" and agricultural economy. Paul Olivier, who runs an engineering firm in Louisiana, has also suggested the plant could be used for biofuel or, at the very least, burned for energy production in the process of eradicating it. There is some scientific research being done on the plant with a view to its potential as a biofuel, with presentations at conferences in 1985 and 2005. The tree has also been determined to be a potential source of wood for construction and particleboard. However, the majority of research currently underway is related to its status as a current pest, not as a potential crop. A list of the current crops being considered for biofuel production in Texas does not include the Chinese tallow tree. Hawaii's discussion of biofuel does not include mention of the tree either.

Biofuel technology has a strong base in Hawaii because the state does need to find alternatives to importing petroleum. Shell has invested in research looking at producing fuel from algae grown and harvested in ponds. Hawaii is a leader in the reuse of cooking oil for transportation fuel, and research has shown that methanol from existing crops is a viable energy alternative that could supply about 10 percent of Hawaii's fuel needs. The waste from sugarcane, pineapple, and other cash crops, as well as eucalyptus and the native leucana tree, are all potential biomass sources under consideration. In fact, electricity has been produced in Hawaii from sugarcane waste biomass as early as 1935. What's more, Hawaii's ecosystem is delicate and unique. Tourism in Hawaii accounts for about a third of the state's gross product, so preventing degradation of the ecosystem is a primary concern for more than one reason. The thought of introducing such a known aggressive invader raises hackles.

Clearly, the idea to use the Chinese tallow tree as a biofuel in Hawaii is not such a great one. However, Mason is not alone in his biofuel judgments; among scientists there is currently a full-fledged debate raging regarding the use of invasive species for biofuels. The issue is that the characteristics of a plant that make it an excellent biofuel candidate are many of the same characteristics found in highly successful destructive invaders. The list of these characters includes: long canopy life, no known pests or diseases, storage of nutrients below ground, rapid growth in spring to outcompete weeds, and efficient use of water. In fact, some of the top candidates for biofuel crops in North America today include invasive species like giant reed, *Miscanthus* hybrids, reed canary grass, and switchgrass. The latter species are native to parts of the United States; however, they can be invasive in other regions. Controlling the spread of these species would be extremely difficult because chemical control is too expensive on rangelands or government lands. Biological control is also likely to be avoided, as there is a risk of the agent spreading beyond the intended species to target other genetically related plants, such as corn, oats, and wheat.

Most exotic species introduced into foreign lands are not

successful. The displaced organisms do not take hold and instead fizzle out and die. It is a very small subset that becomes invasive. Those that do get established, however, tend to do extremely well — and outstrip the local competition. In many ways, inventors such as Stanley Mason that venture outside their field of expertise are themselves an alien species. When they delve into foreign territory and apply their innovative abilities to foreign topics, they are more likely than not to miss the mark. Their ideas, lacking experience and theoretical background, disappear without a trace. Mason's idea for the Chinese tallow tree in Hawaii is one such seed that did not get established.

Every so often, however, an inventor who delves outside his or her field can approach a problem unencumbered by background knowledge and come up with a solution that not only takes hold but thrives, primarily *because* it is exotic. In *The Sources of Invention*, a short list of inventions developed by people working outside their field includes: Gillette, a cork salesman who invented the safety razor; Carlson, a patent lawyer who invented xerography; the undertaker who invented the telephone dialing system; the musicians who invented Kodachrome; and Dunlop, a veterinarian who invented the pneumatic tire.

It is a gamble to venture outside your expertise; you might turn up trumps, but more likely turnips.

AVOID ACCIDENTS

BUCKMINSTER FULLER'S

FISH-SHAPED PEOPLE CARRIER

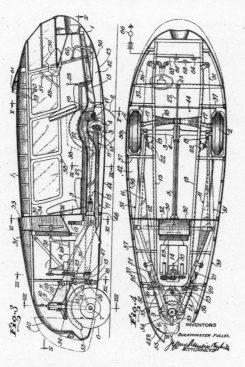

Buckminster Fuller was granted a patent for his Dymaxion car in 1937. More like a ship or a plane than a car, the Dymaxion was fish-shaped, three-wheeled, and had rear steering.

1895–1983

Robert Buckminster Fuller was farsighted. As a young child, he could not see details. Faces were only shadows, and houses, trees, people, and the moon were blurred patterns. It was not until kindergarten that this extreme hyperopia was discovered, and Bucky was given glasses and the ability to see finer features. Farsightedness remained, however, characteristic of Bucky's view of the world throughout his entire life. His thoughts, theories, and inventions were all on a grand scale. He thought in "Whole Systems" using "Total Thinking" and made "Grand Strategy Decisions." He believed that technology could take care of every human on Earth and make humanity "a success in the universe." Buckminster Fuller believed that he knew the key to saving the world: "comprehensive, anticipatory design science." His futuristic, holistic plans are brilliant. They are also unrealistically simple, disregarding smaller but significant details, like cultural preferences.

Fuller was a short, stocky man with a huge intellect. He did not set out to be an inventor or designer. In fact, throughout his life, he defied pigeonholing, being more than the sum of his parts and preferring to be known as a "comprehensive anticipatory design-science" explorer and "a verb." He did, however, identify himself with two groups: he carried around his International Association of Machinists' membership card with pride, and he considered himself a sailor. He was indeed a sailor and spent the First World War in the United States Navy. Fuller's love of the water is reflected in several aquatic inventions, including an innovative means of moving a boat manually, which consists of an umbrellalike folding cone on a pole. Opening and closing the cone moves the boat as though there is a powerful jellyfish behind it, propelling it forward. He also patented a catamaran of sorts, a highly efficient means to row oneself across a body of water with minimal effort.

Bucky's capacity to think about physical forces and three-dimensional space was evident from an early age. In kindergarten, he amazed his teacher with models of tetrahedrons made with toothpicks and dried peas. Another teacher was called in to view the masterpiece. The year was 1899. Needless to say, Bucky did not fit into school very well, with its tidy rows of tidy desks, tidy facts, and tidy minds. As history has it, he took particular offense to geometry lessons. When a point was defined on the blackboard as dimensionless and a line was defined as made up of points, a plane as parallel lines, and a cube as a solid stack of square planes whose edges are equal, young Bucky is said to have put up his hand. "I have some questions. How long has the cube been there? How long is it going to be there? How much does it weigh? And what is its temperature?"

By the evening of his life, world famous and spending most of his days jetting around the globe public speaking, Fuller had not changed. He said, "What usually happens in the educational process is that the faculties are dulled, overloaded, stuffed and paralyzed, so that by the time most people are mature they have lost many of their innate capacities." Perhaps this dislike of the establishment explains why sailing appealed to him so much. A sailor's life depended on the interplay between the physical forces of the elements and the self-contained ship floating among them. Only truth — a genuine understanding of those forces and how to use them — could keep a ship a functioning unit.

Bucky Fuller's nonconformity and refusal to see the world from the down-to-Earth, this-is-the-way-it-is viewpoint led him into difficulty when he followed in the footsteps of four generations of Fullers and attended Harvard University. Bored, he took off to New York, visited Marilyn Miller, an actress he knew, and subsequently took an entire chorus line out to dine at Churchill's — one of New York's finest. He charged the bill to the family bank account. His family, and Harvard, were not happy; he was expelled.

When given a second chance, Fuller was no more enamored with the system partly, it seems, because he objected to the compartmentalization of knowledge — it opposed his holism. He was thrown out almost as soon as he restarted. But it was not for good. As is the way with genius, out from under the "auspices" of any formal organization, Fuller's ideas eventually came to fruition. Forty-eight years later, the degreeless Fuller was back at Harvard.

This time to teach! He held the illustrious post of Charles Eliot Norton Professor of Poetry in 1962. Over the course of his lifetime, he acquired 50 honorary doctorates and 100 major awards of merit, including the Presidential Medal of Freedom.

Fuller's childhood fascination with geometry was an honest interest that became a prevalent aspect of his design work. After not graduating from Harvard, Fuller spent most of the early years of his career working in industry, largely in managerial roles. His first real attempt at marketing an invention was a collaboration with his father-in-law, James Monroe Hewlett, an artist and designer in his own right. The pair had developed a stackable building system — in essence bricks with holes in them. Called the Stockade Building System, a company was formed to manufacture and sell the product, with Fuller as president. The system did away with the need for laborious mortar work. Instead, all the builder needed to do was vertically line up the holes in a wall of bricks then pour cement down them to form inner support columns. Both sides were plastered, and the end result was an 8-inch (20 cm) thick wall, which was fire and moisture proof. The system was cost effective, simple, and strong.

It failed — quite spectacularly. The reasons why are a common theme in Fuller's life and are captured nicely in a comedy skit by the BBC's Mitchell and Webb, which depicts Bronze Orientation Day, when the stone tool workers are told by a traveling educator from the "Tribe in the Valley That Have Lots of Jewelry All of a Sudden" that they'll need to face up to the fact that stone is obsolete and they'll have to retrain or the Tribe with Bronze Axes will kill them. And then take all their stone axes. And "throw them away because they're rubbish." Stackable brick buildings do away with the need for traditional bricks and bricklayers, disrupt the brick manufacturing industry, and even threaten the need for engineers. In 1927, Mr. Hewlett was forced to sell his stock, and Fuller was told by the new owners that his services were no longer needed.

It was 1927, and Fuller found himself disillusioned, unemployed, broke, and seriously contemplating suicide. But, he reasoned, it was his duty to give the world the contents of his mind, gleaned from his own novel combination of intellect and experience. He had "a blind date with principle." Instead of disappearing into Lake Michigan, he locked himself away in his dingy Chicago apartment for a couple of years, thinking and thinking from first principles about the world's problems and how he might fix them.[1] When he emerged, he tried to make a gift of his patents to the American Institute of Architects, who responded by changing their policy so it clearly stated their opposition to all "peas-in-a-pod-like repro- ducible designs." The Institute disliked Fuller's stackable bricks so much that they made it a rule to oppose the entire concept — not just stackable bricks, but any other future designs that could be easily repeated. The Institute felt that architecture should be defined by "one-offs" and only "one-offs."

Years later, the architecture officialdom that had thrown him out invited him back in. In his lifetime, Fuller was awarded a gold medal for Architecture from the National Institute of Arts and Letters, as well as five awards from the American Institute of Architects, though he never identified himself as an architect per se and was never officially qualified as one. It is in architecture, as well as in poetry, that Fuller has achieved lasting fame. Hugh Kenner managed to define his complexity succinctly: "It's a poet's job he does, clarifying the world." All of which is highly descriptive, but it leaves out a driving force in Fuller's psyche, a powerful altruism.

A pamphlet was circulated in 1928, the result of Fuller's post-suicidal contemplations. In much too small a font, using Fuller's best (or worst) convoluted speech, it proclaimed his solutions to the world's problems of poverty, inequality, and the environment. It was called 4-D, after the fourth dimension. In addition to explain-ing some basic laws of the universe, Fuller also outlined his idea for

---

1 Stanford University Press has just published a new book, *New Views on R. Buckminster Fuller* (2009). Based on letters and other papers in their archives, the book's editors, Hsiao-Yun Chu and Roberto Trujillo, claim that the near-suicide story, epiphany, and subsequent silence were a myth created by Fuller to explain the origin of his manifesto and the launching of his career.

a shelter, the 4-D house. The shelter was based on many of Fuller's favorite fundamental design principles. Using minimal materials, cost, and space with maximum utility and efficiency, his shelter did not rely on the weight of its structural materials for strength, but instead it made intelligent use of the opposing forces of tension and compression. The result was a structure that was self-sufficient, whole. It could be airlifted into place (the original plan used a dirigible for the task), allowing humans to live, inexpensively, wherever they desired. It could solve a lot of problems, thought Fuller, including poverty.

In oversimplified terms, the structure resembled a ship's mast, with the rigging holding up multiple hexagonal platforms (the floors of the house). It could be erected in one day and was relatively disaster-proof. While the design of this living space was unique, it was not the full gist of Fuller's plan, for his ambition was to make this ultimate shelter affordable to every person and as accessible as a Model T. It was to be manufactured en masse and put together on site. It was a "peas-in-the-pod-like, reproducible design" taken to extremes. Fuller designed it with the purpose of eliminating all elements of individual design and doing away with the need for home maintenance and even housework. Fuller thought his house would free humanity to do whatever they wished with their time.

Not much later, in 1929, a trademark of sorts was coined to describe Fuller's work: Dymaxion. A department store in Chicago, Marshall Field, wanted to use Fuller's 4-D house as a prop — to help sell their advanced design furniture. But they wanted a better name than 4-D and assigned an advertiser, Waldo Warren, the task of coming up with one. Warren took from Fuller's speech a series of words made from four syllables. Fuller chose from the list the word Dymaxion, a combination of "dynamism," "maximum," and "ions." The word fit, and Marshall Field copyrighted it in Fuller's name.

Fuller's shelter invention, now called the Dymaxion house, came with a whole host of ideas about domestic efficiency. He had ideas for self-sufficient energy units with recycled water, pneumatic doors and beds, and rotating shelves in the walls that delivered what you wanted with the push of a button. There were automatic laundry facilities too. Think *The Jetsons*, yet remember that Fuller had worked out these ideas decades before this space-age cartoon captured 1960s' imaginations. Fuller fully believed this kind of automation and self-sufficient shelter was feasible. Yet he knew that technology was not advanced enough to make it happen on the grand scale required to bring it to the masses. As early as 1927, he estimated that the investment in that technology would be in the realm of a $1 billion. By 1932, he recalculated the capital investment to be $100 million, and in 1945, it was down to $10 million. Once the manufacturing technology was in place, however, Fuller thought he could sell a fully equipped Dymaxion house for about $1,500.

Like many of Fuller's grand ideas, a scaled-down version was produced that was potentially sellable. While we are still waiting for the automated laundry center to arrive in the marketplace, Fuller did produce a Dymaxion bathroom in 1937. This self-contained unit measured 5 feet square (0.5 m²) and came in four pieces. It included a shower/bath, a toilet, a sink, and a mirrored medicine cabinet. The shower sprayed "fog" not water, in keeping with Fuller's observation in the navy that water vapor could clean a man's very dirty face with great efficiency. The bathroom was inexpensive and could be put in place and bolted together by amateurs. A plumber was required only to "plug it in," a task that took a matter of minutes. The bathroom's design was not governed completely by efficiency however — it had several luxurious touches. Water from the sink's single faucet was premixed to just the right temperature, and the faucet sprayed its water toward the back — away from the user's sleeves, forever doing away with wet cuffs. Also, the bathtub had lights beneath the waterline, for dramatic effect.

Twelve Dymaxion bathrooms were made, and some were actually sold. The response from the general public was warm enough; however, rumor has it that manufacturers were afraid of plumbers' unions and wouldn't touch it. The Dymaxion bathroom had the same fate as the stackable building system; it looked like Fuller's ideals of inexpensive and efficient improvements were to be squashed by those that stood to gain from expensive, inefficient lifestyles.

Fuller's ideas for shelters continued to evolve, and there were several designs, all echoing the same principles of minimal materials and cost while maximizing space and utility. The geodesic dome, United States Patent Number 2,682,235, granted in 1954, brought Fuller prestige, official accolades, and income. It came about because of Fuller's explorations in geometry. He did not believe in pi — his argument being that nature does not use an irrational constant in the construction of spheres, such as bubbles. He had a point. Fuller set out to systematically understand and model the true nature behind nature's structures and thus understand the fundamental laws of materials. He made significant progress in uncanny ways that cutting-edge research continues to prove correct.[2] It is out of this "energetic-synergetic geometry" that Fuller developed the geodesic dome.

Geodesic is the term given to any arc of a circle. A geodesic dome is a partial sphere consisting of networks of spherical triangles (i.e., triangles with curved, not straight, sides) formed by geodesics. The domes common in children's playgrounds are of Fuller's design. Triangles are structurally strong, and Fuller's geodesic domes have amazing structural properties, including tremendous stability, excellent volume-to-surface-area ratio (they are comparatively large inside in relation to the area of ground they occupy), and low weight. In fact, the larger the dome, the lighter it is, relative to its volume. Fuller estimated that a mile-wide dome would weigh about the same as the air inside it. He thought that heating the air inside by even one degree would make the sphere airborne. Fuller's imagination had no bounds; he envisioned airborne cities enclosed in such spheres, as well as gigantic domes to cover entire Earth-bound cities. In addition to these structural features, geodesic domes made

---

2 The 1996 Nobel Prize was awarded for the discovery of C60, which has the shape of a "Bucky ball."

of repeated units are inexpensive and quick and easy to assemble. In 1954, a 55-foot (16.8 m) diameter and 40-foot (12.2 m) high dome made of fiberglass was erected in 14 hours; in testing, it remained stable at wind velocities of 220 miles per hour (354 km/h).

Geodesic domes found a market, and many special buildings around the world are evidence of that. The first was a dome to sit atop the Ford Motor Company rotunda, built in 1953. The Montreal Biosphere, the Epcot Center at Disney World, Shaw's Garden in St. Louis, Missouri, and military stations around the world also bear witness to the appeal of the design. There are an estimated 300,000 domes scattered around the globe. Ironically, these days, the geodesic dome is more likely to be considered a 1960s fashion fad than the ultimate in human shelter.

Shelter was not the only human problem Fuller put his mind to solving. He recognized that the common map of the world gives a distorted view. So he developed a Dymaxion map, first published in 1943, which uses triangles to divide the world and provide a much more accurate two-dimensional image of the globe. Fuller is the first person to have held a patent for cartography. The approach allows for greater manipulation of viewpoint, as the triangles can be split up to produce a two-dimensional image in any which way, without distortion. Laid out with the North Pole near the center, we see that the world's continents are not best described in two strips — eastern and western hemispheres with oceans dividing them — but are rather closely connected via the Arctic. It is thus more accurate to say the continents form one very large convoluted landmass. One continent, one ocean: a totally different view of the world indeed, and a more accurate one at that. How very Bucky.

Transportation was on Fuller's list too, and the Dymaxion car originated in his 4-D stage. The original 4-D vehicle was an automobile that could fly. It was aerodynamic, fuel efficient, and the mechanism for changing direction, the steering, was located in the back — like a plane or a boat. Original sketches of the 4-D auto-airplane show inflatable wings and a propeller that would be locked in place during road use. Fuller reasoned that this car/plane could work using "a twin jet stilt" system akin to a duck's wings. The wings of most birds, and planes, are shaped so that a partial vacuum is created above them, producing lift, with the result that they are "sucked" upward and suspended in mid-air. Ducks, however, are too heavy in the body for the size of their wings to make use of this piece of aeronautical physics. Ducks get airborne by pushing hard downward with their outstretched wings, compressing the air in the pockets between their wing and their body, creating jets of down-ward air — twin air stilts — that force them up. Without the tech-nology to make the twin jet stilts and inflatable wings (it was the early 1930s after all), Fuller settled for a v-8 Ford engine and set out to build "the land-taxiing phase of a wingless, twin-orientable-jet-stilts flying device." The vehicle was known to "self-balancing, twenty-eight-jointed adapter-base bipeds" (meaning us) as the Dymaxion car. At least Fuller's fellow citizens would benefit to some degree from his design skill in the present day, even though they'd have to wait some time for technology to catch up to his bigger plans. And make no mistake, this being Fuller, the vehicle was indeed part of a much grander plan:

> I built . . . experimental transports of appropriate aeronautical con-formation, embodying strategic steerability controls in respect to even-tual omni-directionable steerability, loadability and maneuvering as these related to centers of balance, effort and stress. I hoped that these experiments would hasten man's practical realization and enjoy-ment of all-medium navigation by hoverable, yet swift, spot-alighting and spot-take-off transportation, thus opening up vast new ranges of preferred earth-dwellability, when extension of chemistry of the met-allurgical heat-level strength rations appropriate to thrust-supported, plummeting, air-land-water craft should be realized.

The first Dymaxion car was built in 1933. It was 19.5 feet (6 m) long and seated 10 people plus a driver. Its fish shape, reminiscent of a small plane, was aerodynamic. The car had gas mileage of 30 or more miles to the gallon (13 km/L) and could travel 120 miles per hour (193 km/h). It had three wheels: two in front and one in the rear. The rear wheel acted as a rudder and was further designed like a boat or plane in that it used a kingpin and cable steering assembly.

Fuller reasoned that front steering was merely a holdover from horse-drawn carriages. His rear-steering setup allowed for significant maneuverability. Unrestricted by a hub assembly, the Dymaxion could turn 360 degrees with its nose rotating around a one foot (30 cm) diameter circle. To park, the head of the fish was driven into the desired spot, and the tail was flipped around gracefully. Its body was made of aluminum, the chassis was made from chrome-molybdenum aircraft steel, and the windows were the same thin, shatterproof glass used in planes. Unlike ordinary cars, the belly of the car was smooth, with most of the working parts enclosed — there was no exposed muffler or gas tank — other than the wheels of course.

Reaction to this strange-looking vehicle was mixed. Fuller described driving it through the streets of New York with a load of reporters and being stopped by a traffic cop at an intersection. While talking to the fellow, Fuller slowly spun the car around him, so the unfortunate man had to keeping turning to keep face to face. The cop was amazed. So was the onlooking policeman directing traffic at the next intersection,[3] and Fuller was required to repeat the trick when he got there. And so on and so on; it took an hour to drive the mile from 57th Street to Washington Square. Others were afraid of the vehicle and considered it unsafe.

To judge the Dymaxion car and the public's reaction to it, it is prudent to put it in context. The Ford four-door sedan, the 1933 Model 40, seated five people, was 15.5 feet (4.7 m) long, had a square profile, and fuel consumption of about 15 miles per gallon (6 km/L). Its top speed was 78 miles per hour (126 km/h). By comparison, the Dymaxion was right out of this world.

The original 4-D transport concept, that of a car with optional

---

3 There were no traffic lights in 1933.

wings "is possible," says retired NASA Aeronautical Engineer Paul Soderman, "but impractical." As a car, the wings would be heavy, difficult to store and attach, and you'd need to carry around a propeller or jet engine as well as a tail. As a plane, the vehicle would require a higher level of maintenance, and safety issues could arise from things like a ding in a parking lot scuffle. That said, there are numerous examples of hybrid plane/cars in various stages of working models today, including the Transition by Terrifuga and the Autovolnator by Paul Moller.

Judgment of Fuller's Dymaxion car from a modern point of view justifies some of the safety concerns. The streamlining of Fuller's design is impressive for its time. The jury is out, however, on how the Dymaxion might fare in wind tunnel tests compared to modern rivals. Paul Soderman points out some areas that could stand improvement. For example, the nose of the car could be less rounded and more elliptical. The shape of the Dymaxion car does not minimize airflow under the body. In Soderman's words, "The Dymaxion has a nicely rounded underside of the nose, and more importantly is at angle of attack relative to the stern. At high speed the wind would tend to lift the tail off the road with disastrous consequences."

Fuller claimed the car was extremely stable and did not skid in a turn, unlike the front-steered competition. However, cross winds were an issue, one that Fuller himself recognized. He felt it could be overcome with proper driver training. The Dymaxion design contributed to the crosswind problem because the rear of the car was tapered laterally but not vertically as well. This meant that there was about as much car exposed to crosswinds at the back as the front. There was more weight at the front (about 75 percent of the load was above the front wheels), and the two front wheels had better ground contact than the single back wheel — resulting in the back end swinging in the wind. Certainly, by today's standards, safety would be a concern, but that does not mean it was unsafe compared to the competition of the 1930s.

Despite the vehicle's obvious potential both in terms of its practical features and initial public response, the car did not make it past the prototype stage. A group in Britain was interested in the Dymaxion, and the car was about to be put on the world's stage at the World's Fair in Chicago in 1933 when tragedy struck. Driven by a professional race-car driver and with the world famous aviation expert Col. Forbes-Sempill as a passenger, the car was on its way to the fair when they had an accident. The car was overturned, the driver was killed, and Forbes-Sempill sustained major injuries. The newspapers had a field day. The Dymaxion was called a "freak car" and declared hazardous. The British group changed their minds, and the Dymaxion deal was scrapped. Though the press did not report it, the accident was actually caused by a collision with another car, driven by a Chicago South park commissioner, who left the scene. It was the collision, not the Dymaxion's design, that had resulted in the fatal crash.

Subsequent to the Dymaxion's demise, Fuller went on to develop two more Dymaxion prototypes — Dymaxion 2 and Dymaxion 3. Fuller was asked by the New York Automobile Show to exhibit the second vehicle in the Grand Central Palace in 1934. Chrysler had purchased space to display their new Air Flow car and requested that the invitation to Fuller be withdrawn; it was not fair, after all, to allow competition to exhibit for free. In a small, poetic way, justice was served by the New York City Police Chief General Ryan, who provided Fuller with a parking space on the street directly in front of the entrance to the Grand Central Palace. The Dymaxion created quite a stir, and the crowds were enamored with Fuller's live parking performance.

Dymaxion 1 was restored after its initial fatal accident but later perished in a garage fire. Dymaxion 2 is on display at the National Auto Museum in Reno, Nevada, although you cannot see the interior, apparently because it was once used as a chicken coop, and the chickens made a permanent mess of it. Dymaxion 3 has disappeared.

In 1943, Fuller redesigned the vehicle, though this last model was never built. It would have had gas mileage of 50 miles to the gallon (21 km/L), as well as three-wheel independent steering, which would have given the vehicle the capacity to move directly sideways, like a crab, a feature that engineer Paul Soderman says "is cute but way too complicated for the task." Today, there are no cars remotely similar to the Dymaxion. For comparison, a 2008 GMC nine-seater van gets 16 miles to the gallon (7 km/L).

The Dymaxion car begs the question why there are no 10-seater vehicles with gas mileage and top speeds equivalent to Fuller's 1930s vehicles. Cost is definitely one factor; the Dymaxion, for example, was made from aluminum. However, material science has come a long way. And, as Henry Ford pointed out, it is possible for expensive technology to be used in mass production provided the cost is low enough that volume of sales compensates for the decrease in the per-sale profit margin. There are other forces at work influencing what cars are on the market, besides those generated by crosswinds. As Chris Paine's 2006 documentary *Who Killed the Electric Car?* points out, the car industry has an aversion to change simply because it affects profits. Not to suggest that the Dymaxion car's design is good enough, or safe enough, to be on the roads today, but it certainly had enough merit to warrant thorough investigation by car companies.

While the car industry may be particularly adept at disabling the new competition before it can get past the green flag, the plight of the independent inventor as sideliner is likely an inherent part of the job. Generally and historically speaking, the biggest social innovations are the work of independent inventors rather than paid industrial engineers.[4] Fuller recognized this. He once said that all human advances originate "in the outlaw area." It is common sense, really, as industry has its methods, investments in manufacturing processes, and answers to the bottom line. Individuals, on the other hand, are free to think from first principles, outside the box, and outside the boundaries of profits and systems. However, for many reasons, the independent inventor is likely to fail, and perfectly sound inventions are unlikely to be accepted.

---

4 Of course, industry *does* also advance technology and benefit society.

For one, inventors generally possess a suite of creative personality characteristics that does not include financial acuity. For another, the patent system fails to adequately protect independent inventors and arguably makes it more difficult for an independent inventor in some ways, not easier. Companies are sometimes reluctant to pay royalties to an independent inventor, and they would rather look inward to their own, prepaid, R&D team. And companies are generally going to be particularly reluctant to start on a new design paradigm when their business is built around an existing one. In addition, industrial giants always have power that individuals do not wield. The few Thomas Edisons of the world are exceptional, and perhaps as much a product of time and place as personality. It seems that Fuller was wrong with at least one of his primary tenets, that good design is enough. In the case of the Dymaxion Car, good design did not do it. Efficiency, and even public interest, did not do it.

Fate is a reckless driver.

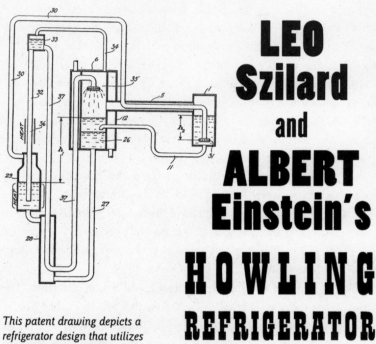

LEO Szilard and ALBERT Einstein's HOWLING REFRIGERATOR

This patent drawing depicts a refrigerator design that utilizes electromagnetic energy to circulate a coolant. Result: a fridge that is safe, but very loud.

• • • • • • • • • • • • • •

1898–1964
and
1879–1955

• • • • • • • • • • • • • •

Baby-faced Leo Szilard was standing in line at an Austro-Hungarian Army camp in Kufstein, on the German border. The year was 1917, and he was a young military officer-in-training. He was feeling really ill, could barely stand up in fact, but did not want to let on because the rule was that no one who was ill could go home. Szilard was worried that he might need a hospital, and he didn't fancy the idea of being stuck in a Kufstein facility, far from his family in Budapest. So, he decided to try to sweet talk his way into a temporary leave of absence on the grounds of a fake emergency at the family home. His plan succeeded, and a weak, feverish Szilard endured the day's journey to Budapest, where he was admitted to hospital. A week later he got a letter from his commanding officer telling him that his class had been sent to the front. Another week passed, and another letter came; this one said the school had been closed down due to a Spanish flu epidemic. Meanwhile, his regiment had come under severe attack on the front, and all had perished. Szilard thought he might have been the first in his school or perhaps the first in the entire Austrian army to have the Spanish flu. He recounts this tale as an example of the importance of timing, and his talent for it.

Young Leo was a rather clever person and a nonconformist. He went to engineering school in Berlin, at the age of 22, because it was a field with potential jobs, but while in Berlin he was drawn to physics and eventually wandered over to the University of Berlin. No small wonder, as the university was a physics hotbed at that time, with Max Planck, Max von Laue, Walter Nerst, Fritz Haber, and Albert Einstein as teachers.

In Germany, a PhD was earned when a graduate student worked on a problem they'd thought up, or one that was assigned by a professor, and presented a paper on their results. If the dissertation was deemed worthy, a PhD was granted. Courses were not graded; students would merely receive a signature at the end of term by the instructor to indicate they had attended. Szilard the PhD student went to von Laue, who gave him a problem to work on. But Szilard was not able to make headway with the topic. Instead, he came up with his own question, in an unrelated field — thermodynamics. Within three weeks he had a manuscript. He was not sure if it was any good, and he was embarrassed because he had not done the work he had been told to do. So, he took it to Professor Einstein and told him what he'd done. Herr Professor said, "That's impossible. This is something that cannot be done" — then he read it. Szilard was right; the professor was impressed, and so a lifelong association had begun, which evolved from student and teacher to business partners in 1926, all because of the refrigerator.

The first decades of the 20th century were full of change in lifestyle, as the waves of the second industrial revolution continued to ripple through society. One of the major changes in everyday life brought about by the advent of wide-scale electricity was mechanical refrigeration. Keeping food cold throughout the summer months has always been a major challenge for civilizations, with direct consequences for health and survival. For millennia, people have been using ice for cold storage, and in the late 1800s, ice was still the major method of cooling for industrial and home use. However, there were issues — such as contamination with sewage. Mechanical cooling was first applied to industry, including meat packing and brewing, in the 1890s. It was only a matter of time before mechanical cooling was doled out to the masses; the first home-cooling units were introduced to the United States between 1911 and 1920, and by 1921 there were 5,000 refrigerators manufactured for home use.

All the many refrigerator models available in the early decades of the 20th century worked on the same principle; as the gas phase of matter expands, it takes up heat from the environment and converts thermal energy to other forms of energy (e.g., kinetic or chemical, depending on the scenario). The same principle is in operation when rubbing alcohol on your skin evaporates, cooling your skin as it does so. In refrigerators, a gas is compressed and under pressure is changed to liquid. The liquid is forced through coils, where it vaporizes, removing heat from the surrounding environment (i.e., from inside the refrigerator). A pump that is run by a motor sucks up the vapor, compresses it into liquid again, and sends it through the system for another bout of cooling. The cooling happens inside the sealed refrigerator; the conversion to liquid with its associated release of heat happens outside the refrigerator.

In the early days of refrigerators, in the 1920s, the gases used in refrigerators were methyl chloride, sulphur dioxide, or ammonia — all of which are toxic, particularly methyl chloride. A single refrigerator contained a lethal dose of that gas. If the gas leaked into a family home, the result was typically fatal.[1] A report published in *The Journal of the Royal Society for the Promotion of Health* in 1930 by a Chicago Department of Health official reveals 10 fatalities related to refrigerators, as well as 185 cases of illness and 879 cases of people driven from their homes because of refrigerant gases. An inspection of refrigerator systems revealed 472 leaks. The article goes on to describe a typical case, where a family of two adults and five children were taken to hospital with symptoms of poisoning; two of the children died.

Years later, Szilard told colleague Bernard Feld that it was just such a family tragedy that inspired Einstein in the winter of 1925–1926 to approach Szilard with the idea of forming a partnership to invent a safe means of refrigeration. A few months later they signed a written agreement with the following terms:

- All inventions in the field of refrigeration would be joint property.
- Szilard would have first claim on profits if his income fell too low.
- Otherwise, all royalties would be shared equally.

It is not so strange as you might think for these giants on the physics stage to delve into invention. Einstein's uncle was an inven-

---

1 A methyl chloride leak from a refrigerator in 1942 was the cause of a fire in the Boston nightclub, the Cocoanut Grove, which led to 492 fatalities.

tor who developed an electric generator and an electricity meter. Einstein had worked in the Swiss Patent Office upon completion of his dissertation after he failed to get an academic job in physics; his job was to review patent applications. He also advised the new graduate Szilard to take up a job at the patent office: "Why don't you take a job in the Patent Office? That would be best for you; it is not a good thing for a scientist to be dependent on laying golden eggs. When I worked in the Patent Office, that was my best time of all." And Einstein kept his hand in invention; the refrigerator was not just a one-off. A hearing aid,[2] a gyroscope, and a camera[3] are among his creations.

As for Szilard, his initial interest and training in engineering were a reflection of his mechanical aptitudes, and he was an inventor as well as a productive scientist. However, he did not carry any of his inventions through to commercial production. Szilard's ideas include several particle accelerators, such as the cyclotron and betatron, and the electron microscope. His first patent was for an X-ray sensitive cell, granted in 1923. He also made numerous contributions to the development of atomic energy and held numerous patents in this field for reactors and associated equipment. He is also known as the "reluctant father of the atomic bomb." His first patent in that area, upon his discovery of a means to produce a nuclear chain reaction, was taken out in 1934[4] and assigned to the British Admiralty, in the hopes of controlling the use of this technology for good and preventing its use for evil. Near the end of his life, Szilard developed cancer, and with his physician wife he formulated a radiation therapy that he used on himself with success. Szilard was a highly conscientious man, as his many actions

---

2 German Patent Number 590783.

3 US Patent Number 2,058,562.

4 British Patents 440,023 and 630,726.

during the development of atomic energy show.[5] He had a philosophical bent, which is reflected in his creation of his own ten commandments:

1. Recognize the relationships between things and the laws which govern men's actions, so that you know what you are doing.

2. Direct your deeds to a worthy goal, but do not ask if they will achieve the goal; let them be models and examples rather than means to an end.

3. Speak to all others as you do to yourself, without regard to the effect you make, so that you do not expel them from your world and in your isolation lose sight of the meaning of life and the perfection of the creation.

4. Do not destroy what you cannot create.

5. (Untranslatable pun.)[6]

6. Do not desire what you cannot have.

7. Do not lie without need.

8. Honor children. Listen to their words with reverence and speak to them with endless love.

9. Do your work for six years; but in the seventh, go into solitude or among strangers, so that the memory of your friends does not prevent you from being what you have become.

10. Lead your life with a gentle hand and be ready to depart whenever you are called.

---

5 A large portion of Szilard's life was dedicated to nuclear physics. As his understanding and that of his colleagues advanced to the point that a bomb seemed possible, Szilard, with typical foresight, fought hard to have the results of the research kept secret from the brewing powers in continental Europe. He failed. He then turned his energies to convincing the American government that in a world in which Nazi Germany was working on an A-bomb, they needed to get there first. In no small way, Szilard contributed to the American success in achieving that goal. Szilard also fought hard against the United States government's plan to actually use the bomb he'd help to build. His entire life, Szilard actively promoted world peace and means to use science for the good of mankind.

6 A version of Szilard's commandments on the Internet by Dr. Ray Cooper translates this as "Touch no dish unless you are hungry," a pun that could read "Do not turn to the court of law unless you are hungry."

The two physicist inventors, Szilard and Einstein, put together their knowledge of thermodynamics, and their big hearts, to develop numerous refrigerator designs that were safe. Eventually, they settled on three designs that represented alternate methods of achieving cooling in a refrigerator based on different principles: absorption, diffusion, and electromagnetism. All of these had one thing in common — the absence of moving parts, for it was this feature that the dynamic duo considered to be the root of the refrigerant leak problem. Moving parts require bearings and seals, and these things inevitably wear out. The pair wasted no time, and in the fall of 1926, the patent filing began. There were to be 45 refrigerator-related patents in all.

One early design was an improvement on an Electrolux model already on the market, which they considered to be the best available. This model was one of a group that used heat from a flame (e.g., gas or propane) to drive the cooling system, rather than a piston. Heat from a flame drove the expansion of the gas that lead to the transfer of heat from the air to the coils. Einstein and Szilard felt this so-called absorption refrigerator, and others like it, were safer. In their new and improved design, a heat source drove butane through a series of chambers and tubes. Pressure was decreased by the presence of ammonia to reduce the boiling point of the butane. In a chamber in the presence of ammonia, the butane vaporized, drawing heat from its surroundings. The gaseous mixture then traveled to another chamber that had water in it. The water dissolved the ammonia, and the now liquid butane was recirculated.

In 1926, Szilard began approaching companies with their absorption refrigerator design, and he instigated work on the building of prototypes at the Institute of Technology in Berlin. There were three early customers. One company, Bamag-Meguin, did not remain in the venture long due to external issues. A Swiss arm of Electrolux bought a patent for an absorption refrigerator for the equivalent of $750, about $10,000 in today's dollars, and all parties thought they had a good bargain. As it turns out, Electrolux did not act on the patent; it appears instead that they were eliminating the competition. Later, Electrolux also bought a patent for a diffusion model; this was not acted upon either. Protecting their own designs likely motivated that purchase as well. A patent for versions of the absorption refrigerator was granted in the United Kingdom in 1928 and another in the United States in 1930.[7]

Another design, of Einstein's genius, was powered by running water — specifically, the pressure of water from a water tap. The pressure produced a vacuum in a chamber in which water and methanol were evaporated. The evaporation cooled the device, which could then be immersed in, say, a glass of water, which would in turn be cooled. The device was simple, portable, and not very expensive. The methanol, however, was slowly consumed with use, so the device would have needed to be replaced periodically. A German company, Citogel, was quite keen on the idea. However, it did not pan out, as the water pressure in German taps varied from house to house and floor to floor. This variation made it impossible to develop a device that would work for everyone.

The third design, the electromagnetic pump, operated on the same principle as the regular mechanical refrigerator models on the market, then and now, except that it was not a piston that compressed the refrigerant, it was a metallic fluid. The fluid was made to move through the tube, like a piston, by an electromagnetic field. This pump worked well, and it had no mechanical moving parts. About two years from the first patent filing, the German General Electric Company (AEG) came on board. They had their own research laboratories, and a team of three, led by Szilard, was assembled. Patent royalties and consultancy fees lined Szilard's pockets, money he used at a later date to keep himself afloat when he fled Nazi Germany — and to finance the initial experiments into nuclear chain reaction.

There were plenty of bugs to be worked out to achieve a working model based on the electromagnetic pump, and the steps are

---

7 US Patent Number 1,781,541.

demonstrative of the inventive process, with one problem's solution creating yet more problems, driving a design in one direction over another. Originally, the design operated via a conduction pump; the liquid metal was made to flow by passing an electric current through it. For this the brainy duo needed a conductive metal. Szilard came up with the idea of a mixture of potassium and sodium, whose alloy is liquid. However, potassium sodium alloy is also highly reactive, and this caused problems in their refrigerator, as it was likely to destroy the insulation on the wires. The solution settled on was to switch from a conduction pump to an induction pump and do away with the need for wires altogether. Instead, the force required to move the metallic fluid was generated with a magnetic field. Being highly reactive, the potassium sodium also posed a risk of explosion if exposed to oxygen. Safety was ensured by enclosing the alloy in a stainless-steel capsule welded shut to make it immune to air exposure. The resulting cooling pump was much less efficient than regular refrigerators, but it never failed, and it never leaked.

The electromagnetic refrigerator had another problem. It howled. Dennis Gabor, a fellow physicist in Berlin, said it "howled like a jackal," while another colleague thought it sounded more like a "banshee." The engineer that worked on all these refrigerators, Albert Korodi, was kinder. He said it sounded like "rushing water." Needless to say, all involved knew the machine was too loud; customers would not want it in their homes, no matter how safe it was. In 1960, Szilard recollected that the refrigerator "was not very practical, because mechanical refrigerators which have moving parts function really quite well and are not too noisy." Work continued in 1931, but the prototype was still not good enough for the open market. In addition to the noise, there was also the problem of its inefficiency.

Work on the Szilard/Einstein refrigerator continued until 1932, with some progress made. A means was found to maintain high enough temperatures within the pump to use pure potassium. This almost doubled the pump's efficiency, though it was still low in comparison to the mechanical pumps.

Several factors combined to settle the matter in favor of the mechanical pump and put the Szilard/Einstein refrigerator into cold storage. The Depression shut down the experimental work of AEG, and manufacturers approached by Szilard in other countries did not bite. With the increasing popularity of the Nazi party, the threat of political unrest was also looming, disrupting industry and the activity of the inventors themselves. In another display of his excellent timing, Szilard saw the signs of trouble in Germany early and transferred all his money from his German bank account to a Swiss one in 1930. He also had his suitcases packed, ready to leave at a moment's notice. When the House of Parliament in Berlin was set ablaze on February 27, 1933, Szilard knew the moment had come. A few days later he was on a train to Vienna never to return to continental Europe. He remembered, "The train was empty. The same train on the next day was overcrowded, was stopped at the frontier, the people had to get out, and everybody was interrogated by the Nazis."[8]

While the world was on the cusp of massive violence, there was another event that sealed the fate of the howling refrigerator; someone else beat them to it. The problem of toxic refrigerants was being solved, elsewhere, in the laboratories of the United States. Throughout the 1920s, refrigerators in the United States had also been causing deaths, sometimes of whole families. Chemists had been working on the problem, and in 1928 they developed a new gas, a chlorofluorocarbon, called Freon for short, coinvented by the great inventor Charles Kettering,[9] who was working for Frigidaire at the time. In a demonstration for the American Chemical Society, Freon was breathed in through a tube and then blown onto a candle. The demonstrator showed no deleterious effects and the

---

8 For a fascinating look at the interplay between science and politics see Weart, S.R. and G. Weiss Szilard, eds., *Leo Szilard: His Version of the Facts* (Cambridge, MA: The MIT Press, 1978). The book provides a sense of the complex issues surrounding science, with relevance for today as well as a unique historical view of the Second World War. Szilard's wit and intellect are reflected in his words as well as his exemplary ethics.

9 Charles Kettering had over 300 patents, including leaded gasoline and the electric starting motor.

candle was extinguished, thus proving its nontoxic and nonflammable properties. Freon quickly transformed the refrigerator market and has been the refrigerant of choice ever since. Today, the refrigerator is America's favorite appliance, found in 99.5 percent of homes.

While the Szilard/Einstein refrigerator was rejected, the Einstein/Szilard electromagnetic pump found an important use in the cooling systems of nuclear reactors. Cool opinions regarding the Szilard/Einstein refrigerators may be warming, as new problems with refrigeration have arisen. One issue is that chlorofluorocarbons (CFCs), including Freon, may not be directly toxic to humans, but they are toxic to the ozone layer that protects our planet. The substances have been banned since 1996, and new refrigerants include hydrofluorocarbons (HFCs), still known collectively as Freon, that are not ozone depleting. These new-age Freons are, however, powerful greenhouse gases. That's one problem.

Another is the lingering safety concern. While not nearly as toxic as methyl chloride, Freon in high concentrations can still be deadly, as a 2008 accident on board a Russian submarine that killed 20 can attest. (The Freon was part of a fire extinguishing system, not a refrigerating one.) Fatalities from accidental leaks from refrigerators still happen, occasionally, though it is fridges from the pre-1960s that are the main culprits.[10]

Yet another refrigeration challenge is how to get cooling capacity without electricity. This line of thinking has two impetuses: green refrigeration that doesn't expend valuable energy or harm the environment and refrigeration for the developing world. Two groups, both working out of Oxbridge in the UK, have reverted back to the Szilard/Einstein designs as a starting point. Malcom McCullough at Oxford is taking another look at the absorption fridge — the butane model. His work is focusing on increasing its efficiency by altering the gases used. This, he says, could quadruple the efficiency. He also

---

10 Ammonia in high concentrations can cause health problems, such as might occur when a leak happens in a small, enclosed space, like an RV; RV refrigerators are generally absorption refrigerators.

hopes to replace the source of heat for the absorption fridge, which in the Szilard/Einstein model, and all other absorption models, has been a fossil-fuel flame. McCullough hopes to make his absorption fridge solar powered. The result would be green, efficient, and low maintenance — ideal for rural applications.

The Cambridge competition is a new company, Camfridge Ltd., whose management team is comprised of two senior engineers and some business associates. Their approach is the electromagnetic howling refrigerator of Szilard/Einstein fame, with some improvements. They promise their gas-free designs provide:

- 50 percent reduction in energy consumption (referring to air conditioning specifically),
- Elimination of maintenance costs as a result of refrigerant leakage,
- Lower cost components,
- And, interestingly, reduction of noise.

Both types of fridges are in research phase and not yet ready for the marketplace, but the race is on for a better refrigerator. The contestants would do well to heed Szilard: "If you want to succeed in the world, you don't have to be much cleverer than other people. You just have to be one day earlier. This is all it takes."

Many references and notes in this book refer to websites. The complete URL where the material was originally accessed online has been included. However, interested readers are invited to www. edisonsconcretepiano.com, where a complete list of the URLs referenced is provided as direct links. More information and photos can be found at this website as well.

In addition, activities and exercises to accompany the book, for use at home or in the classroom, are available for free download at www.edisonsconcretepiano.com.

Introduction

Page 13

Kongshem, L., "Face to Face: Alan Kay Still Waiting for the Revolution," *Administr@tors Magazine* (April/May 2003), www2.scholastic.com/browse/article.jsp?id=5.

Wolfe, T., "Land of Wizards," *The Best American Essays 1987*, ed. G. Talese (New York: Ticknor and Fields, 1987).

Page 14

Gladwell, M., *Outliers: The Story of Success* (New York: Little, Brown and Company, 2008).

Jewkes, J., D. Sawers, and R. Stillerman, *The Sources of Invention*, 2nd ed. (London: Macmillan and Co. Ltd., 1969) 82.

Page 16

Rossman, J., *The Psychology of the Inventor* (Washington, DC: The Inventors Publishing Co., 1931) 2.

Weick, C.W. and C.F. Eakin, "Independent Inventors and Innovation: An Empirical Study," *International Journal of Entrepreneurship and Innovation* 6, no. 1 (2005): 5–15.

Saturday Evening Post, September 27, 1930.

Page 17

Schmookler, J., *Invention and Economic Growth* (Cambridge, MA: Harvard University Press, 1966) 25.

Page 18

Petroski, H., *Design Paradigms: Case Histories of Error and Judgment in Engineering* (Cambridge, MA: Cambridge University Press, 1994) 1–2.

Leonardo da Vinci's Walk on Water Shoes

Page 22

Bramly, S., *Leonardo: Discovering the Life of Leonardo da Vinci* (New York: HarperCollins, 1991) 396.

Vasari, G., *Lives of the Artists*, trans. George Bull. Harmondsworth (1965), quoted in S. Bramly, *Leonardo: Discovering the Life of Leonardo da Vinci*, (New York: HarperCollins, 1991) 4.

Page 23

Freud, S., *Eine Kindheitserinnerung des Leonardo da Vinci* [Leonardo Da Vinci: A Memory of His Childhood] (London: Routledge, 1984) 88.

Bramly, S., *Leonardo: Discovering the Life of Leonardo da Vinci* (New York: HarperCollins, 1991) 36–37.

da Vinci, Leonardo, *Codex Atlanticus*, vol. 327, quoted in M. Kemp, *Leonardo da Vinci: The Marvellous Works of Nature and Man*, (Oxford: Oxford University Press, 2006) 82.

Page 24

da Vinci, Leonardo, "MS. A.," fol. 47r, quoted in I.B. Hart, *The Mechanical Investigations of Leonardo da Vinci* (London: Chapman and Hall Ltd., 1925) 78.

Truesdell, C.A., "The Mechanics of Leonardo da Vinci," in *Essays in the History of Mechanics* (New York: Springer-Verlag, 1968) 8.

Kemp, M., *Leonardo* (Oxford: Oxford University Press, 2004) 38–40.

Bramly, S., *Leonardo: Discovering the Life of Leonardo da Vinci* (New York: HarperCollins, 1991) 220–222.

Burke, J., "Meaning and Crisis in the Early Sixteenth Century: Interpreting Leonardo's Lion," *Oxford Art Journal* 29, no. 1 (2006): 77–91.

Page 25

Kemp, M., *Leonardo da Vinci: The Marvellous Works of Nature and Man* (Oxford:

Oxford University Press, 2006) 30–35, 76–78, 208–209.

Clark, K., *Leonardo da Vinci*, 4th ed. (London, UK: The Folio Society, 2005) 42, 60–61.

Museo d'Arte e Scienza, Milan, www.leonardodavincimilano.com.

Page 26

Kemp, M., *Leonardo* (Oxford: Oxford University Press, 2004) 10–46, 95–97, 109–112, 199–200.

Bramly, S., *Leonardo: Discovering the Life of Leonardo da Vinci* (New York HarperCollins, 1991) 297, 384–385.

Page 27

Bramly, S. *Leonardo: Discovering the Life of Leonardo da Vinci* (New York: HarperCollins, 1991) 179, 234–240.

da Vinci, Leonardo, *Codex Atlanticus*, fol. 117 b, quoted in I.B. Hart, *The Mechanical Investigations of Leonardo da Vinci*, (London: Chapman and Hall Ltd., 1925) 43.

Kemp, M., *Leonardo* (Oxford: Oxford University Press, 2004) 119.

Vasari, G., *Lives of the Artists*, trans. George Bull. Harmondsworth (1965), quoted in S. Bramly, *Leonardo: Discovering the Life of Leonardo da Vinci* (New York: HarperCollins, 1991) 5.

Page 28

Clark, K., *Leonardo da Vinci*, 4th ed. (London: The Folio Society, 2005) 149.

Kemp, M., *Leonardo* (Oxford: Oxford University Press, 2004) 116–122.

We do know he owned a copy of *De re militari* by Roberto Valturio. Bramly, S., *Leonardo: Discovering the Life of Leonardo da Vinci* (New York: HarperCollins, 1991) 270.

Bramly, S., *Leonardo: Discovering the Life of Leonardo da Vinci* (New York: HarperCollins, 1991) 324–329, 341.

Page 29

Kemp, M., *Leonardo* (Oxford: Oxford University Press, 2004) 30, 127.

Kemp, M., *Leonardo da Vinci: The Marvellous Works of Nature and Man* (Oxford: Oxford University Press, 2006) 98–100.

Bramly, S., *Leonardo: Discovering the Life of Leonardo da Vinci*. (New York: HarperCollins, 1991) 313–326.

Clark, K., *Leonardo da Vinci*, 4th ed. (London: The Folio Society, 2005) 182.

Page 30

Vasari, G., *Lives of the Artists*, trans. George Bull. Harmondsworth (1965), quoted in S. Bramly, *Leonardo: Discovering the Life of Leonardo da Vinci*, (New York: HarperCollins, 1991) 388.

di Giorgio Martini, Francesco, *Trattato di Architectura*, (c. 1475) Cod. 148, f. 47v, quoted in Truesdell, C.A., "The Mechanics of Leonardo da Vinci," in *Essays in the History of Mechanics* (New York: Springer-Verlag, 1968) 5.

Page 31

Roach, J., "How 'Jesus Lizards' Walk on Water," *National Geographic News*, (November 16, 2004), news.nationalgeographic.com/news/2004/11/1116_041116_jesus_lizard_2.html.

Riordan, T., "Patents: An Inventor Traces the Footsteps of Leonardo da Vinci with a Device That Walks on Water," *The New York Times* (August 2, 2004), query.nytimes.com/gst/fullpage.html?res=9C0DEEDA1F3DF931A3575BC0A962 9C8B63.

Page 32

British Council, "Water Trivia," Learn English Central, www.britishcouncil.org/ learnenglish-central-trivia-water.htm

Philipkoski, K., "Get Out and Walk. It's Only Water," *Wired* (April 14, 2003), www.wired.com/medtech/health/news/2003/04/58457

Page 33

Kemp, M., *Leonardo* (Oxford: Oxford University Press, 2004) 2, 48.

Page 34

Kemp, M., *Leonardo da Vinci: The Marvellous Works of Nature and Man* (Oxford: Oxford University Press, 2006) 67.

da Vinci, Leonardo, quoted in J.P. Richter and R.C. Bell, eds., *The Notebooks of Leonardo da Vinci* (Mineola, NY: Courier Dover Publications, 1970) 796.

da Vinci, Leonardo, *Quaderni III*, 12v quoted in S. Bramly, *Leonardo: Discovering the Life of Leonardo da Vinci* (New York: HarperCollins, 1991) 376.

Vasari, G., *Lives of the Artists*, trans. George Bull. Harmondsworth (1965), quoted in S. Bramly, *Leonardo: Discovering the Life of Leonardo da Vinci* (New York: HarperCollins, 1991) 407.

Vasari, G., *Lives of the Artists*, trans. Elizabeth L. Seeley (1885), /www.efn.org/~acd/vite/VasariLeo.html.

Randall, J.H. Jr., "The Place of Leonardo da Vinci in the Emergence of Modern Science," in *The School of Padua and the Emergence of Modern Science* (Padova, Italy: Editrice Antenore, 1961) 115–138.

Truesdell, C.A., "The Mechanics of Leonardo da Vinci," in *Essays in the History of Mechanics* (New York: Springer-Verlag, 1968) 1–83.

Page 35

da Vinci, Leonardo, (British Museum, 180b), quoted in S. Bramly, *Leonardo: Discovering the Life of Leonardo da Vinci* (New York: HarperCollins, 1991) 401.

James Watt's Apparatus for the Administration of Medicinal Airs

Page 37

How James Watt's apparatus worked: Fire is started inside the furnace by putting coal in through the side door of the furnace (1). The side door into the cast iron fire tube (2) has a round opening to permit the addition of starting materials, like charcoal or slacked lime. Secondary materials, such as water, are inserted through the top funnel (3). The airs given off are transmitted to the oiled silk bag (5) via the long tube (4) possibly made of tin. A patient would inhale the contents of the silk bag through the mouthpiece (6), which is stopped by a wooden plug.

Page 39

Dickinson, H.W., *James Watt: Craftsman and Engineer* (New York: Augustus M. Kelley Publishers, 1967) 117–118.

Rolt, L.T.C., *James Watt* (New York: Arco Publishing Company, Inc., 1962) 22, 133.

Arago, M., *Historical Eloge of James Watt* (New York: Arno Press, 1975) 97, 123.

Carnegie, A., *James Watt* (New York: DoubleDay, Page and Co., 1905) 149.

Uglow, J., *The Lunar Men* (London: Faber and Faber Ltd., 2002) 104.

Page 40

Rolt, L.T.C., *James Watt* (New York: Arco Publishing Company, Inc., 1962) 46.

Dickinson, H.W., *James Watt: Craftsman and Engineer* (New York: Augustus M. Kelley Publishers, 1967) 79–83.

Watt to Joseph Black, January 13, 1779, in *Partners in Science: Letters of James Watt and Joseph Black*, eds. E. Robinson, and D. McKie (Cambridge, MA: Harvard University Press, 1970) 50.

Watt to Joseph Black, October 25, 1780, in *Partners in Science: Letters of James Watt and Joseph Black*, eds. E. Robinson and D. McKie (Cambridge, MA: Harvard University Press, 1970) 99.

Arago, M., *Historical Eloge of James Watt* (New York: Arno Press, 1975) 176.

Page 41

Dickinson, H.W., *James Watt: Craftsman and Engineer* (New York: Augustus M. Kelley Publishers, 1967) 79, 114.

Rolt, L.T.C., *James Watt* (New York: Arco Publishing Company, Inc., 1962) 17–18.

Uglow, J., *The Lunar Men* (London: Faber and Faber Ltd., 2002) 97.

Page 42

Watt to Joseph Black, May 5, 1794, in *Partners in Science: Letters of James Watt and Joseph Black*, eds. E. Robinson and D. McKie (Cambridge, MA: Harvard University Press, 1970) 202.

Watt to Joseph Black, April 7, 1788, in *Partners in Science: Letters of James Watt and Joseph Black*, eds. E. Robinson and D. McKie (Cambridge, MA: Harvard University Press, 1970) 165.

Page 43

Watt to Joseph Black, June 6, 1784, in *Partners in Science: Letters of James Watt and Joseph Black*, eds. E. Robinson and D. McKie (Cambridge, MA: Harvard

University Press, 1970) 143.

Watt to Joseph Black, May 21, 1798, in *Partners in Science*, eds. E. Robinson and D. McKie (Cambridge, MA: Harvard University Press, 1970) 293.

Rolt, L.T.C., *James Watt* (New York: Arco Publishing Company, Inc., 1962) 109.

Page 44

Revolutionary Players, www.search.revolutionaryplayers.org.uk/engine/resource/exhibition/standard/default.asp?theme=234&originator=%2Fengine%2Fcustom%2Fplace.asp&page=&records=&direction=&pointer=113&text=0&resource=709.

Jones, J., H.S. Torrens, and E. Robinson, "The Correspondence Between James Hutton (1726–1797) and James Watt (1736–1819) with Two Letters from Hutton to George Clerk-Maxwell (1715–1784): Part II," *Annals of Science* 52 (1995): 357–382.

Watt to Joseph Black, January 7, 1796, in *Partners in Science: Letters of James Watt and Joseph Black*, eds. E. Robinson and D. McKie (Cambridge, MA: Harvard University Press, 1970) 221–222.

Page 45

Jones, J., H.S. Torrens, and E. Robinson, "The Correspondence Between James Hutton (1726–1797) and James Watt (1736–1819) with Two Letters from Hutton to George Clerk-Maxwell (1715–1784): Part II," *Annals of Science* 52 (1995): 357–382.

Watt to Joseph Black, August 31, 1794, in *Partners in Science: Letters of James Watt and Joseph Black*, eds. E. Robinson and D. McKie (Cambridge, MA: Harvard University Press, 1970) 207.

Page 46

Jones, J., H.S. Torrens, and E. Robinson, "The Correspondence Between James Hutton (1726–1797) and James Watt (1736–1819) with Two Letters from Hutton to George Clerk-Maxwell (1715–1784): Part II," *Annals of Science* 52 (1995): 357–382.

Watt to Joseph Black, April 2, 1795, in *Partners in Science: Letters of James Watt and Joseph Black*, eds E. Robinson and D. McKie (Cambridge, MA: Harvard University Press, 1970) 214.

Watt to Joseph Black, October 13, 1796, in *Partners in Science: Letters of James Watt and Joseph Black*, eds E. Robinson and D. McKie (Cambridge, MA: Harvard University Press, 1970) 228.

Watt to Joseph Black, November 10, 1799, in *Partners in Science: Letters of James Watt and Joseph Black*, eds. E. Robinson and D. McKie (Cambridge, MA: Harvard University Press, 1970) 310.

Jonker, M.A., "Maximum Likelihood Estimation of Life-span Based on Censored and Passively Registered Historical Data," *Lifetime Data Analysis*, 9 (2003): 35–36.

Page 47

Watt to Joseph Black, November 22, 1799, in *Partners in Science: Letters of James Watt and Joseph Black*, eds. E. Robinson and D. McKie (Cambridge, MA: Harvard University Press, 1970) 312

Jones, J., H.S. Torrens, and E. Robinson, "The Correspondence Between James Huttom (1726–1797) and James Watt (1736–1819) with Two Letters from Hutton to George Clerk-Maxwell (1715–1784): Part II," *Annals of Science* 52 (1995): 357–382.

Robert Hooke's Flying Machine Powered by Artificial Muscle

Page 49

Hooke claimed to have made successful working models of flying machines based on endless screws, windmills, gunpowder, and springs. However, he did not write down how he made these models. Whether the models represented in this drawing might actually fly or not is not known.

Fig. 1

A: Windmill turns with the wind.

B: Gear is driven off turning endless screw, gear alternately pushes then pulls rod fixed to wing.

C: Wing is fixed to frame and is moved up and down by the rod attached to the rotating gear.

D: Handle "manages" the contraption by engaging and disengaging the gear against the turning, windmill-driven endless screw.

Operation: The machine is fitted to a chair or to a person's back, and then turned into a stiff wind, with the inclined windmill directly facing the wind. The windmill responds by rotating, the base of the windmill shaft has an endless screw that meets a gear. The gear has a dowel fitted snugly into its periphery and a rod fitted to it. The other end of the rod is connected to the wing at its mid point. As the gear is turned the rod end moves up and down. The flying device is stopped by operating the lever (D), which disengages the gear from the endless screw.

Fig. 2 A later model

A: Cylinder with piston; cylinder is fixed to frame and piston is fixed to top side of wing.

B: Gunpowder is kept in a reservoir.

C: A tension spring pulls wing up after piston has discharged.

D: Wing is attached to frame and to piston.

E: A close gate valve is attached to the wing to allow for the opening of the chamber in order to deliver the gunpowder when the wing returns to "up position."

Operation: A piston (A) has a small oil torch fixed to its side. When the wing is
pulled upwards by the spring, the tension on the spring on the gunpowder
valve (E) opens, allowing gunpowder to spill into the cylinder. This promptly
ignites, pushing the piston down and closing a valve (E) at the end of its
extension, once the explosive power of the powder is exhausted. The spring
(C) pulls the wing back up and the process repeats.

Page 52

Hooke, R., *Micrographia* (1665; New York: Dover Publications, Inc., 1961) preface.

Purrington, Robert D, *The First Professional Scientist: Robert Hooke and the Royal
Society of London*, vol. 39, Science Networks/Historical Studies (Basel,
Switzerland: Birkhäuser, 2009).

Page 55

Inwood, S., *The Man Who Knew Too Much: The Strange and Inventive Life of
Robert Hooke, 1635–1703* (London: Pan Macmillan Ltd., 2002) 132.

Page 56

Inwood, S., *The Man Who Knew Too Much: The Strange and Inventive Life of
Robert Hooke, 1635–1703* (London: Pan Macmillan Ltd., 2002) 237–238.

Hooke, Robert, diary entry dated Friday, June 2, 1676, in H.W. Robinson and W.
Adams, eds., *The Diary of Robert Hooke, 1672–1680* (London: Taylor and
Francis, 1935) 235.

Page 57

Hooke, Robert, June 26, 1689, quoted in S. Inwood, *The Man Who Knew Too
Much: The Strange and Inventive Life of Robert Hooke, 1635–1703* (London:
Pan Macmillan Ltd., 2002) 2.

Inwood, S., *The Man Who Knew Too Much: The Strange and Inventive Life of
Robert Hooke, 1635–1703* (London: Pan Macmillan Ltd., 2002) 23–24.

Page 58–59

Inwood, S., *The Man Who Knew Too Much: The Strange and Inventive Life of
Robert Hooke, 1635–1703*. (London: Pan Macmillan Ltd., 2002) 115, 177, 185,
213.

Page 60

Chapman, A., *England's Leonardo: Robert Hooke and the Seventeenth Century
Scientific Revolution* (Bristol and Philadelphia: Institute of Physics Publishing,
2005).

Cooper, M., *A More Beautiful City: Robert Hooke and the Rebuilding of London
After the Great Fire* (London: Sutton Publishing, 2003).

Cooper, M. and M. Hunter, *Robert Hooke: Tercentennial Studies* (Farnham, UK:
Ashgate Publishing Co., 2006).

Cooper, M., *Robert Hooke and the Rebuilding of London* (Stroud, UK: Sutton
Publishing Ltd., 2005).

Jardine, L., *The Curious Life of Robert Hooke: The Man Who Measured London*

(New York: HarperCollins, 2003).

Bennett, J., M. Cooper, M. Hunter, and L. Jardine, *London's Leonardo: The Life and Work of Robert Hooke* (Oxford: Oxford University Press, 2003).

May, R.M., "Address by Lord May PRS, given at the dedication of the memorial to Dr. Robert Hooke at Westminster Abbey on 3 March 2005," *Notes & Records of the Royal Society* 59 (2005): 321–323.

Page 61

E.g., Baskett, T.F. "Robert Hooke and the Origins of Artificial Respiration." *Resuscitation* 60 (2004): 125–127; Gest, H. "The remarkable Vision of Robert Hooke (1635–1703): First Observer of the Microbial World." *Perspectives in Biology and Medicine* 48 (2005): 266–272; Harsch, V. "Robert Hooke, Inventor of the Vacuum Pump and the First Altitude Chamber (1671)." *Aviation Space and Environmental Medicine* 77 (2006): 867–869; Hintzman, D.L. "Robert Hooke's Model of Memory." *Psychonomic Bulletin & Review* 10 (2003): 3–14; Hunter, M. "Robert Hooke Revivified." *Notes & Records of the Royal Society of London* 58 (2004): 89–91; Lim, M.V.V. "The History of Extracorporeal Oxygenators." *Anaesthesia* 61 (2006): 984–995; McConnell, A. "Origins of the Marine Barometer." *Annals of Science* 62 (2005): 83–101; Mills, A. "Robert Hooke's 'Universal Joint' and Its Application to Sundials and the Sundial-clock." *Notes & Records of the Royal Society of London* 61 (2007): 219–236; Nauenberg, M. "Hooke and Newton: 'Divining' Planetary Motions." *Physics Today* (February 2004): 13–14; Nauenberg, M. "Robert Hooke's Seminal Contribution to Orbital Dynamics." *Physics in Perspective* 7 (2005): 4–34; Rousseaux, G., P. Coullet, and J.M. Gilli "Robert Hooke's Conical Pendulum from the Modern Viewpoint of Amplitude Equations and Its Optical Analogues." *Proceedings of the Royal Society of London* 462 (2006): 531–540; Stevenson, C. "Robert Hooke, Monuments and Memory." *Art History* 28 (2005): 43–73.

Inwood, S. *The Man Who Knew Too Much: The Strange and Inventive Life of Robert Hooke, 1635–1703* (London: Pan Macmillan Ltd., 2002) 42.

Page 62

Drake, E.T., *Restless Genius: Robert Hooke and His Earthly Thoughts* (New York: Oxford University Press, 1996) 71, 98.

Waller, R., *The Life of Dr. Robert Hooke: Posthumous Works* (1705; New York: Johnson Publishing, 1969).

Page 63

Inwood, S. *The Man Who Knew Too Much: The Strange and Inventive Life of Robert Hooke, 1635–1703* (London: Pan Macmillan Ltd., 2002) 111.

Hart, C., *The Prehistory of Flight* (Los Angeles: University of California Press, 1985).

Page 64

Wilkins, J., *Mathematicall Magick* (London, 1648).

Waller, R., *The Posthumous Works of Robert Hooke, M.D., S.R.S.* (1705) in R.T. Gunther, *The Life and Work of Robert Hooke*, Early Science in Oxford, vol. 6, (Oxford: Oxford University Press, 1930) 9–10.

Page 65

Gunther, R.T., *The Life and Work of Robert Hooke*, Early Science in Oxford, vol. 6 (Oxford: Oxford University Press, 1930) 44, 293.

Hooke, Robert, diary entries dated: October 4, 1674, October 7, 1674, October 8, 1674, February 11, 1675, December 23, 1675, December 23, 1675, December 28, 1675, January 11, 1676, March 30, 1676, October 24, 1677, May 31, 1679, in H.W. Robinson and W. Adams, eds., *The Diary of Robert Hooke* (London: Taylor and Francis, 1935).

Page 66

Gunther, R.T., *The Life and Work of Robert Hooke*, Early Science in Oxford, vol. 6 (1930) 362.

Poore, S., "The Morphological Basis of the Arm-to-Wing Transition," *Journal of Hand Surgery* 33 (2008): 277–280.

Powell, D., "Were Pterosaurs Too Big to Fly?" *New Scientist*, (October 2, 2008), www.newscientist.com/article/mg20026763.800-were-pterosaurs-too-big-to-fly.html?feedId=dinosaurs_rss20.

Page 68

Gibbs-Smith, C.H., *Aviation: An Historical Survey from Its Origins to the End of World War II* (London: Science Museum, 1970).

www.jet-man.com

Page 69

The American Helicopter Society Igor I. Sikorsky Human Powered Helicopter Competition, vtol.org/awards/hph.html.

Hart, C., *The Prehistory of Flight* (Los Angeles: University of California Press, 1985) 197, 200.

Hooke, Robert, diary entry dated Friday, May 19, 1676, in H.W. Robinson and W. Adams, eds., *The Diary of Robert Hooke* (London: Taylor and Francis, 1935) 233.

Thomas Edison's Concrete Piano

Page 73

Although Edison's ideas for a concrete piano did not get off the drawing board, and his application for a patent for a method of producing concrete furniture was rejected, the Lauter Piano Company brought his idea to life in 1931. In their patented piano, a mixture of cement and sawdust was poured into molds. The resulting concrete piano looked like any other baby grand, but was a little heavier.

Page 75

Edison, T.A., *The Diary and Sundry Observations of Thomas Alva Edison*, ed. D.D. Runes, (New York: Philosophical Library, 1948) 45.

Page 76

Clark, R.W., *Edison: The Man Who Made the Future* (New York: G.P. Putnam's Sons, 1977) 31.

Page 77

Edison, T.A., *The Diary and Sundry Observations of Thomas Alva Edison*, ed. D.D. Runes (New York: Philosophical Library, 1948) 179.

Jonnes, J., *Empires of Light: Edison, Tesla, Westinghouse, and the Race to Electrify the World* (New York: Random House, 2003) 148–150.

Clark, R.W., *Edison: The Man Who Made the Future* (New York: G.P. Putnam's Sons, 1977) 69, 101.

Page 78

Clark, R.W., *Edison: The Man Who Made the Future* (New York: G.P. Putnam's Sons, 1977) 61, 204, 231.

Page 79

Clark, R.W., *Edison: The Man Who Made the Future* (New York: G.P. Putnam's Sons, 1977) 89.

A complete list of Edison's patents is available at www.tomedison.org/patent.html.

Page 81

Quoted in: Gelb, M.J. and S. Miller Caldicott, *Innovate Like Edison: The Success System of America's Greatest Inventor* (New York: Dutton Books, 2007) 37.

Page 83

Punch (October 30, 1907), quoted in R.W. Clark, *Edison: The Man Who Made the Future* (New York: G.P. Putnam's Sons, 1977) 185.

Edison, T.A., *The Diary and Sundry Observations of Thomas Alva Edison*, ed. D.D. Runes (New York: Philosophical Library, 1948) 20, 82, 168.

Wachhorst, W., *Thomas Alva Edison: An American Myth* (Cambridge, MA: MIT Press, 1981) 16.

Page 84

Edison, T.A., *The Diary and Sundry Observations of Thomas Alva Edison*, ed. D.D. Runes. (New York: Philosophical Library, 1948) 86.

Primary Abbreviated Case File Title Concrete Furniture, fol. 821, ser. 674274 (Thomas A. Edison Papers, Rutgers University).

For an image of the cabinet described see Clark, R.W. *Edison: The Man Who Made the Future* (New York: G.P. Putnam's Sons, 1977) 199.

This and other models can be viewed at: www.nps.gov/edis/photosmultimedia/photogallery.htm?eid=87885&aid=221&root_aid=210.

Page 86

Hunter, Kim, telephone interview with author, August 2007.

Mechanical Music Digest,
    www.mmdigest.com/Archives/Digests/200404/2004.04.13.07.html.
Page 87
Email correspondence with Joe Wolfe, November 2007.
Sauter, E., "Tilt-up Provides Building Solutions for Modern churches," *Concrete Engineering International* 10 (2006): 57–59.
Page 88
Clare, M. and W. Ward, "Moving forward," *Concrete Construction* 51 (2006): 26–34.
Page 89
Clark, R.W., *Edison: The Man Who Made the Future* (New York: G.P. Putnam's Sons, 1977) 25, 211.
Edison, T.A., *The Diary and Sundry Observations of Thomas Alva Edison*, ed. D.D. Runes (New York: Philosophical Library, 1948) 205–244.
Page 90
Edison, T.A., *The Diary and Sundry Observations of Thomas Alva Edison*, ed. D.D. Runes (New York: Philosophical Library, 1948) 5, 43, 65, 163.

Nikola Tesla's Earthquake Machine

Page 93
In the patent for his reciprocating engine Tesla writes, "In the invention which forms the subject of my present application, my object has been, primarily to provide an engine, which under the influence of an applied force such as the elastic tension of steam or gas under pressure will yield an oscillatory movement which, within very wide limits, will be of constant period, irrespective of variations of load, frictional losses and other factors which in all ordinary engines produce change in the rate of reciprocation." The machine was used both to start an earthquake, as well as to build a vibrating platform.
Page 94–95
Cheney, M., *Tesla: Man out of Time* (Englewood Cliffs, NJ: Prentice-Hall, Inc., 1981) 2.
McGovern, C., "The New Wizard of the West," *Pearson's Magazine* (May 1899).
O'Neill, J.J., *Prodigal Genius: The Life of Nikola Tesla* (New York: Ives Washburn, Inc., 1944) 154–158.
Page 96–97
Sparling, E., "Nikola Tesla, at 79, Uses Earth to Transmit Signals; Expects to Have $100,000,000 Within Two Years," *New York World Telegram*, July 11, 1935.
O'Neill, J.J., *Prodigal Genius: The Life of Nikola Tesla* (New York: Ives Washburn, Inc., 1944) 159–162.

Page 98

The Tacoma Narrows Bridge collapse was captured on film and can be viewed
online at www.archive.org/details/Pa2096Tacoma.

For more info on the *Mythbusters* earthquake machine episode, see Wikipedia:
en.wikipedia.org/wiki/MythBusters_%28season_4%29#Episode_60_.E2.80.94_
.22Earthquake_Machine.22.

Page 99

Jonnes, J., Empires of Light (New York: Random House, 2003) 354–362.

Cheney, M., Tesla: Man out of Time. (Englewood Cliffs, NJ: Prentice-Hall, Inc.,
1981) 33.

Page 100

For more info on Powercast see www.powercastco.com.

Tesla, N., Chapter 3 in *My Inventions: The Autobiography of Nikola Tesla*,
www.lucidcafe.com/library/96jul/teslaauto03.html.

Email correspondence with Chathan Cooke, December 2007.

Page 101

Tesla, N., Chapter 1 in *My Inventions: The Autobiography of Nikola Tesla*,
www.lucidcafe.com/library/96jul/teslaauto06.html.

Sacks, O., *An Anthropologist on Mars* (New York: Vintage Books, 1995).

Page 102

Tesla, N. Chapter 3 in *My Inventions: The Autobiography of Nikola Tesla*,
www.lucidcafe.com/library/96jul/teslaauto03.html.

Page 104

Stanford University, "Stanford Researchers Establish a Link Between Creative
Genius and Mental Illness," *Science Daily* (May 22, 2002),
www.sciencedaily.com/releases/2002/05/020522073047.htm.

Tesla, N., "Man's Greatest Achievements," in J.J. O'Neill, *Prodigal Genius: The Life
of Nikola Tesla* (New York: Ives Washburn, Inc., 1944) 251–252.

Page 105

Cheney, M., *Tesla: Man out of Time* (Englewood Cliffs, NJ: Prentice-Hall, Inc.,
1981) xiii.

Norman, R., *A Pictorial Tour of Unarius* (Unarius, CA: 1982).

The *Tesla Speaks* books are available for purchase online at: www.unarius.org.

Tesla Memorial Society of New York, "'To Mars with Tesla; or, the Mystery of
Hidden Worlds,' A Science Fiction Tale from 1901, Tesla and the Exploration of
Cosmos," www.teslasociety.com/marswithtesla.htm.

Page 106

For an example of a company that sells Tesla shields see Life Technology,
www.lifetechnology.co.uk/teslashield.htm. There are others.

Henry Ford's Flipping Fordson

Pages 110–111

Burlingame, R., *Henry Ford: A Great Life in Brief* (New York: Alfred A. Knopf, 1954) 30, 56, 74.

Ford, H., *My Life and Work* (Garden City, NY: Doubleday, Page and Co., 1926) 57.

Graves, R.H., *The Triumph of an Idea: The Story of Henry Ford* (Garden City, NY: Doubleday, Doran and Co., Inc., 1934) 46–47.

Bryan, F.R., *Clara: Mrs. Henry Ford* (Detroit: Wayne State University Press, 2001) 143.

Page 112

Arnald, H.L. and F.L. Faurote, *Ford Methods and the Ford Shops* (New York: Arno Press, 1972).

Burlingame, R., *Henry Ford: A Great Life in Brief* (New York: Alfred A. Knopf, 1954) 114.

Ford, H., *My Life and Work* (Garden City, NY: Doubleday, Page and Co., 1926) 108.

Page 113

Ford, H., *My Life and Work* (Garden City, NY: Doubleday, Page and Co., 1926) 98, 130.

Greenleaf, W., *From These Beginnings: The Early Philanthropies of Henry and Edsel Ford, 1911–1936* (Detroit: Wayne State University Press, 1964) 29–69.

Page 114

Greenleaf, W., *From These Beginnings: The Early Philanthropies of Henry and Edsel Ford, 1911–1936* (Detroit: Wayne State University Press, 1964) 18–21, 114.

Ford, H., *My Life and Work* (Garden City, NY: Doubleday, Page and Co., 1926) 133, 188–193, 210–214.

Page 115

Burlingame, R., *Henry Ford: A Great Life in Brief* (New York: Alfred A. Knopf, 1954) 19, 25.

Ford, H., *My Life and Work* (Garden City, NY: Doubleday, Page and Co., 1926) 15.

Page 116

Ford, H., *My Life and Work* (Garden City, NY: Doubleday, Page and Co., 1926) 22–26.

Nevins, A., *Ford: The Times, the Man, the Company* (New York: Charles Scribner's Sons, 1954) 73.

Brendan, Gill, "To Spare the Obedient Beast," *New Yorker* 22 (May 18, 1946): 34, quoted in Nevins, A., *Ford: The Times, the Man, the Company* (New York: Charles Scribner's Sons, 1954) 160.

Page 117

SSB Tractor, "History of Ford Farm Tractors," www.ssbtractor.com/features/Ford_tractors.html.

Includes photos and details of all the models.

Page 118

Ford, H., *My Life and Work* (Garden City, NY: Doubleday, Page and Co., 1926) 202–204.

Page 119

Yesterday's Tractor forum, www.ytmag.com/cgi-bin/viewit.cgi?bd=ttalk&th=682383.

American Society of Agricultural Engineers, "Why 540?" Bnet, October 2004, find-articles.com/p/articles/mi_qa5409/is_/ai_n21357618.

Page 120

Peterson, C. and R. Beemer, *Ford N Series Tractors* (Minneapolis: MBI Publishing Co., 1997) 25.

Page 121

Yesterday's Tractor forum, www.ytmag.com/cgi-bin/viewit.cgi?bd=ttalk&th=682538.

Ford, H., *My Life and Work* (Garden City, NY: Doubleday, Page and Co., 1926) 3, 12–13, 19, 270.

Ford, H., *My Life and Work* (Garden City, NY: Doubleday, Page and Co., 1926) 16, 18.

Page 122

Gordon, J.S., *American Heritage Magazine*, 43, no. 2 (1992), www.americanher-itage.com/articles/magazine/ah/1992/2/1992_2_14.shtml.

Ford, H., *My Life and Work* (Garden City, NY: Doubleday, Page and Co., 1926) 195.

Burlingame, R., *Henry Ford: A Great Life in Brief* (New York: Alfred A. Knopf, 1954) 123.

George Washington Carver's Miracle Peanut Cure

Page 123

Series VII.1, Photographs, Box 7.1/3, file II. Photographs—Carver, George Washington, USDA History Collection, Special Collections, National Agricultural Library.

Page 124

Mackintosh, B., "George Washington Carver: The Making of a Myth," *The Journal of Southern History*, 42, no. 4 (1976): 507–528.

Holt, R., *George Washingon Carver: An American Biography* (Garden City, NY: DoubleDay, Doran and Co., Inc., 1943) 101.

McMurry, L.O., *George Washington Carver: Scientist & Symbol* (New York: Oxford University Press, 1981).

Page 125

George Washington Carver to Booker T. Washington, 1896, in Holt, R., *George*

Washingon Carver: An American Biography (Garden City, NY: DoubleDay, Doran and Co., Inc., 1943) 97.
Correspondence in response to letter from Washington asking Carver to join Tuskegee.
Page 126
Iowa State University, Biographical note preceding a list of George Washington Carver papers, www.lib.iastate.edu/arch/rgrp/21-7-2.html.
Iowa State University, "The Legacy of George Washington Carver" (1998), www.lib.iastate.edu/spcl/gwc/resources/products.html.
McMurry, L.O., *George Washington Carver: Scientist & Symbol* (New York: Oxford University Press, 1981) 154.
Page 128
McMurry, L.O., *George Washington Carver: Scientist & Symbol* (New York: Oxford University Press, 1981) 179, 193.
Morson, A. *Operative Chymist* (Amsterdam: Rodopi, 1997) 217.
Page 129
Agency for Toxic Substances & Disease Registry, Department of Health and Human Services, "Public Health Statement: Creosote," (2002), www.atsdr.cdc.gov/toxprofiles/phs85.html#bookmark05.
Summit Industries, "Creomulsion," www.creomulsion.com.
King, J.C., J. Blumberg, and L. Ingwesen, "Tree Nuts and Peanuts as Components of a Healthy Diet," *The Journal of Nutrition* 138 (2008): 1736S–1740S.
Carsten, R.E., A.M. Bachand, S.M. Bailey, and R.L. Ullrich, "Resveratrol Reduces Radiation-Induced Chromosome Aberration Frequencies in Mouse Bone Marrow Cells," *Radiation Research* 169 (2008): 633–638.
Hurst, W.J., J.A. Glinski, K.B. Miller, J. Apgar, M.H. Davey, and D.A. Stuart, "Survey of the Trans-resveratrol and Trans-piceid Content of Cocoa-containing and Chocolate Products," *Journal of Agricultural and Food Chemistry* 56 (2008): 8,374–8, 378.
Pages 130–131
Jenkins, D.J.A., F.B. Hu, and L.C. Tapsell, "Possible Benefit of Nuts in Type 2 Diabetes," *The Journal of Nutrition* 138 (2008): 1,752S–1,756S.
Mattes, R.D., P.M. Kris-Etherton, and G.D. Foster, "Impact of Peanuts and Tree Nuts on Body Weight and Healthy Weight Loss in Adults," *The Journal of Nutrition* 138 (2008): 1,741S–1,745S.
Traoret, C.J., P. Lokko, A.C.R.F. Cruz, C.G. Oliveira, N. Costa, J. Bressan, R.C.G. Alfenas, and R.D. Mattes, "Peanut Digestion and Energy Balance," *International Journal of Obesity* 32 (2008): 322–328.
Rocha-González, H.I., M. Ambriz-Tututi, and V. Granados-Soto, "Resveratrol: A Natural Compound with Pharmacological Potential in Neurodegenerative Diseases," *CNS: Neuroscience & Therapeutics* 14, no. 3 (2008): 234–247.

McMurry, L.O., *George Washington Carver: Scientist & Symbol* (New York: Oxford University Press, 1981) 84, 214.

Alexander Graham Bell's Six-Nippled Sheep

Page 134

Graham Bell, A., "Sheep-Breeding Experiments on Beinn Bhreagh," *Science* 36, no. 925 (1912): 378–384.

Graham Bell, M., quoted in A. Graham Bell, "Saving the Six-Nippled Breed," *Journal of Heredity* 14 (1923): 99–111.

Page 135

Eber, D.H., *Genius at Work* (Toronto: McClelland and Stewart Ltd., 1982) 14, 20, 81.

Page 136

Eber, D.H., *Genius at Work* (Toronto: McClelland and Stewart Ltd., 1982) 37.

Graham Bell, M., quoted in A. Graham Bell, "Saving the Six-Nippled Breed," *Journal of Heredity* 14 (1923): 99–111.

Page 137

Graham Bell, A., "The Multi-Nippled Sheep of Beinn Bhreagh," *Science* 19, no. 489 (1904): 767–768.

Page 138

Graham Bell, A., "Sheep-Breeding Experiments on Beinn Bhreagh," *Science* 36, no. 925 (1912): 378–384.

Graham Bell, A., "Saving the Six-Nippled Breed," *Journal of Heredity* 14 (1923): 99–111.

Page 139

Bruce, R.V., *Bell.* (Ithaca, NY: Cornell University Press, 1990) 415–416.

"Alexander Bell's Sheep," *Time*, February 9, 1942, www.time.com/time/magazine/article/0,9171,777628-1,00.html.

Page 140

Oftedal, O.T., "The Mammary Gland and Its Origin During Synapsid Evolution," *Journal of Mammary Gland Biology and Neoplasia* 7, no. 3 (2002): 225–252.

Yoon, C.K., "Of Breasts, Behaviour, and the Size of Litters," *New York Times*, October 19, 1999, query.nytimes.com/gst/fullpage.html?res=9505EFD81739F93AA25753C1A96F958260&sec=&spon=&pagewanted=print.

Speert, H., "Supernumerary Mammae, with Special Reference to the Rhesus Monkey," *The Quarterly Review of Biology* 17, no. 1 (1942): 59–68.

Page 141

Buss, D.H., and J.E. Hamner III, "Papio cynocephalus" [Supernumerary nipples in the baboon] *Folia Primatologica* 16 (1971): 153–158.

Goertzen, B.L. and H.L. Ibsen, "Spernumerary Mammae in Guinea Pigs," *Journal of Heredity*, 42 (1951): 307–311.

Derocher, A.E., "Supernumerary Mammae and Nipples in the Polar Bear *(Ursus maritimus)*," *Journal of Mammalogy* 71 (1990): 236–237.

Erickson, A.W., "Supernumerary Mammae in the Black Bear," *Journal of Mammology* 41 (1960): 409.

Pages 142–143

Mackenzie, C., *Alexander Graham Bell* (Boston: Houghton Mifflin Company, 1928) 6–7, 282.

Davis, G.H., "DNA Tests in Prolific Sheep from Eight Countries Provide New Evidence on Origin of the Booroola (FecB) Mutation," *Biology of Reproduction* 66 (2002): 1,869–1,874.

Tamarack Lamb & Wool, www.tamaracksheep.com/booroola.html.

For information and photos of sheep with the booroola gene.

Elihu Thomson's Quartz Telescope Mirror

Page 145

This photo comes from the Elihu Thomson Papers at the American Philosophical Society in Philadelphia, Pennsylvania. There are actually two captions on the back. The other says, "Prof. Thomson inspecting a 22-inch fused-quartz astronomical mirror — one of his experiments leading toward the construction of a 200-inch mirror for the California Institute of Technology telescope at the Mt Wilson observatory."

Page 146

Compton, K.T., *Biographical Memoir of Elihu Thomson, 1853–1937*, Biographical Memoirs, vol. 21, 4th memoir, (Washington, DC: National Academy of Sciences, 1939).

Woodbury, D.O., *Beloved Scientist: Elihu Thomson, a Guiding Spirit of the Electrical Age* (New York: McGraw-Hill Book Company, Inc., 1944) 114.

Carlson, W.B., "Invention, Science, and Business: The Professional Career of Elihu Thomson, 1870–1900" (PhD diss., University of Pennsylvania, 1984).

Page 147

Woodbury, D.O., *Beloved Scientist: Elihu Thomson, a Guiding Spirit of the Electrical Age*, (New York: McGraw-Hill Book Company, Inc., 1944) xi, 159, 213–214.

Pages 148–149

Carlson, W.B., "Invention, Science, and Business: The professional Career of Elihu Thomson, 1870–1900" (PhD diss., University of Pennsylvania, 1984) 146–147.

Woodbury, D.O., *Beloved Scientist: Elihu Thomson, a Guiding Spirit of the Electrical Age*, (New York McGraw-Hill Book Company, Inc., 1944) 13–14, 254.
Page 150
Compton, K.T., *Biographical Memoir of Elihu Thomson, 1853–1937*, Biographical Memoirs, vol. 21, 4th memoir (Washington, DC: National Academy of Sciences, 1939).
Page 151
Woodbury, D.O., *Beloved Scientist: Elihu Thomson, a Guiding Spirit of the Electrical Age* (New York: McGraw-Hill Book Company, Inc., 1944) 259.
Abrahams, H.J. and M.B. Savin, eds., *Selections from the Scientific Correspondence of Elihu Thomson* (Cambridge, MA: MIT Press, 1971) 256, 287.
Pages 152–153
Woodbury, D.O., *Beloved Scientist: Elihu Thomson, a Guiding Spirit of the Electrical Age*. (New York: McGraw-Hill Book Company, Inc., 1944) 216, 328.
Stanford University, einstein.stanford.edu/.
For more information and photos of the Gravity Probe satellite
National Astronomical Observatory of Japan, www.naoj.org.
Space Today, www.spacetoday.org/Japan/Japan/Astronomy.html.
For information and photos of the Subaru telescope.

Danny Hillis's Paint Can Robot

Page 157
In this design drawing of nine-year-old Danny's paint-can robot, the artist has included the following features:
Fig. 1
1. Large garbage can for body.
2. Gallon paint can for head, fitted with Christmas lights for eyes and a speaker from an old radio for a mouth.
3. Rotisserie motor from a barbeque operates both arms.
4. Power supply to operate electronics.
5. Two quart paint cans riveted together for arms.
Fig. 2
1. Three wheels let into the bottom, one at front, two at rear.
2. Rear wheels are run off the power supply.
Pages 158–159
Much of the information in this chapter was gleaned through a phone interview between Danny Hillis and the author in December 2008.

Hillis, W.D., *The Connection Machine* (Cambridge, MA: MIT Press, 1985).
The Long Now Foundation, www.longnow.org.

Hillis, W.D., "The Millennium Clock," *WIRED*, Scenarios ed. (1995), www.wired.com/wired/scenarios/clock.html.

Page 161

Hillis, W.D., "A Forebrain for the Mind World," Edge World Question Center, 2009, www.edge.org/q2009/q09_12.html#hillis.

See the US Patent and Trademark Office, website for a list. Search under "inventor" for Hillis, W Daniel or see: patft.uspto.gov/netacgi/nph-Parser?Sect1=PTO2&Sect2=HITOFF&p=1&u=%2Fnetahtml%2FPTO%2Fsearch-adv.htm&r=0&f=S&l=50&d=PTXT&Query=IN%2F%22hillis%2C+w.+daniel%22

Hillis, W.D., *The Connection Machine*, (Cambridge, MA: MIT Press, 1985) 3.

Page 163

"At Home With Robots: The Coming Revolution," *TechNewsWorld* (November 26, 2008), www.technewsworld.com/story/65290.html.

For a snapshot of the current status of robots in the home.

McCord, M., "At Hong Kong High-tech Café, Everything Is Served with Microchips," *Space Daily*, (October 23, 2006), www.spacedaily.com/reports/At_Hong_Kong_High_Tech_Cafe_Everything_Is_Served_With_Microchips_999.html.

Hong, C., "Robots May Force Chefs Out of the Kitchen," *China Daily* (October 18, 2006), www.chinadaily.com.cn/china/2006-10/18/content_710686.htm .asimo.honda.com.

Page 164

Humanoid Robotics Group, MIT Artificial Intelligence Laboratory, www.ai.mit.edu/projects/humanoid-robotics-group/coco/future-work.html

Descriptions of the work currently underway by the MIT humanoid robotics group.

"Robots Could Demand Legal Rights," BBC News (December 21, 2006), news.bbc.co.uk/2/hi/technology/6200005.stm.

Brockman, J., "Edge annual question, 2009," Edge World Question Center, 2009, www.edge.org/q2009/q09_index.html.

Perkowitz, S., *Digital People: From Bionic Humans to Androids* (Washington, DC: Joseph Henry Press, 2004) 199–219.

Shuster, F., "Survival of the Cutest: Disney Lets Walk-Alone Robot Dinosaur, the First of Its Kind, Out to Play This Week," *Los Angeles Daily News* (August 30, 2003), www.thefreelibrary.com/SURVIVAL+OF+THE+CUTEST+DISNEY+LETS+WALK-ALONE+ROBOT+DINOSAUR,+THE...-a0107145107.

Mouse Planet, www.mouseplanet.com/parkupdates/dlr/dlr030825.htm.

For pictures and videos of Lucky.

www.robosaurus.com.

Page 165

Thompson, C., "It's Alive!" *WIRED*, issue 15.01 (January 2007), www.wired.com/wired/archive/15.01/alive_pr.html

J. Walter Christie's Flying Tank

Page 169

Pittman, W., "The Los Angeles Fire Department's Nine Mechanical Horses," Los Angeles Fire Department Historical Archive, 1988, www.lafire.com/fire_apparatus/1913-1923_tractors/1913-1923_tractors_pittman.htm.

Georgano, G.N., *The Beaulieu Encyclopedia of the Automobile*, reprint ed. (London: Taylor and Francis, 2000).

"History of the Tank," www.globalsecurity.org/military/systems/ground/tank-history1.htm.

Page 170

Hofmann, G.F., "A Yankee Inventor and the Military Establishment: The Christie Tank Controversy," *Military Affairs* 39 (1975): 12–18.

Alexander. A., *Armor Development in the Soviet Union and the United States. A Report for Director of Net Assessment, Office of the Secretary of Defense*, R-1860-NA (Santa Monica, CA: Rand, September 1976) 275–276.

Page 171

Hoffman, G.F. and D.A. Starry, *Camp Colt to Desert Storm: The History of the U.S. Armored Forces* (Lexington, KY: University of Kentucky Press, 1999) 80.
www.geocities.com/firefly1002000/christanks.html.
For a list of Christie's tank models with descriptions.

Hoffman, G.F., *Through Mobility We Conquer: The Mechanization of U.S. Cavalry* (Lexington, KY: University Press of Kentucky, 2006) 63.

Benson, C.C., "The New Christie, Model 1940," *Army Ordnance*, September/October 1929, 112.

Page 172

Hofmann, G., "Doctrine, Tank Technology, and Execution: I.A. Khalepskii and the Red Army's Fulfillment of Deep Operations," *The Journal of Slavic Military Studies* 9 (1996): 288–327.

Alexander. A., *Armor Development in the Soviet Union and the United States. A report for Director of Net Assessment, Office of the Secretary of Defense*, R-1860-NA (Santa Monica, CA: Rand, September 1976) 73.

Page 173

Hofmann, G., "Doctrine, Tank Technology, and Execution: I.A. Khalepskii and the Red Army's Fulfillment of Deep Operations," *The Journal of Slavic Military Studies* 9 (1996): 288–327.

Gillie, M.H., *Forging the Thunderbolt: History of the U.S. Army's Armored Forces, 1917–45* (Mechanicsburg, PA: Stackpole Books, 2006) 34.

Hoffman, G.F., *Through Mobility We Conquer: The Mechanization of U.S. Cavalry* (Lexington, KY: University Press of Kentucky, 2006) 164.

Page 174

Alexander. A. *Armor Development in the Soviet Union and the United States. A*

report for Director of Net Assessment, Office of the Secretary of Defense, R-
1860-NA (Santa Monica, CA: Rand, September 1976) 72.

Gillie, M.H., Forging the Thunderbolt: History of the U.S. Army's Armored Forces,
1917–45. (Mechanicsburg, PA: Stackpole Books, 2006) 33–35.

Page 175

Holt, L., "Flying Tanks That Shed Their Wings," Modern Mechanics and
Inventions, July 1932, 34–37, 167, 170, 176.

Page 176

Hofmann, G., "Doctrine, Tank Technology, and Execution: I.A. Khalepskii and the
Red Army's Fulfillment of Deep Operations," The Journal of Slavic Military
Studies 9 (1996): 288–327.

Holt, L., "Flying Tanks That Shed Their Wings," Modern Mechanics and
Inventions, July 1932, 34–37,167, 170, 176.

Hoffman, G.F., Through Mobility We Conquer: The Mechanization of U.S. Cavalry
(Lexington, KY: University Press of Kentucky, 2006) 152.

Page 177

mailer.fsu.edu/~akirk/tanks/japan/japan-exp.html.
   For an image of the winged tank.

www.aviastar.org/air/england/ga_hamilcar.php.
   For an image of the General Aircraft Hamilcar.

Page 180

ca.youtube.com/watch?v=uQGJh6MyByO&feature=related.
   For YouTube footage of the BMD-4 ACV.

www.globalsecurity.org/military/world/russia/bmd-4.htm.
   For more information on Russian ACVs.

George Davison's Popcorn Volcano

Pages 186–187

A. Sostek, "With a Pirate Ship, Cave and Tree House as Offices, These Designers
May Never Come Home," Pittsburgh Post-Gazette, Thursday, October 26,
2006, www.post-gazette.com/pg/06299/733053-28.stm.

Theodore, D., "The Science of Space," Canadian Architect (March 2005),
www.cdnarchitect.com/issues/ISarticle.asp?id=161750&story_id=63849140810
&issue=03012005&PC=.

Davison, G., G. Davison, Official Website of George Davison, October 2006,
www.georgemdavison.com/2006/10/.

Page 188

Anderson, C., The Long Tail: Why the Future of Business Is Selling Less of More
(New York: Hyperion Books, 2006).

Chris Anderson is *WIRED*'s editor in chief.
Page 189
Hasbro, www.hasbro.com/games/kid-
    games/monopoly/default.cfm?page=History/history.
For more information about Monopoly.
"Artists Are Rebels Regardless of the Cause, Personality Study Shows," (press
    release, University of Melbourne, Melbourne, Australia, October 19, 2005),
    uninews.unimelb.edu.au/articleid_2897.html.
Moore, A.D. *Invention, Discovery, and Creativity* (New York: DoubleDay and Co.,
    1969) 46–52.

Jerome Lemelson's Flying Balloon

Page 192
Davidson, M., "Jerome Lemelson, Independent Inventor (1923–1997)," the
    Lemelson Center for the Study of Invention and Innovation, Smithsonian
    Institution, Washington, DC,
    invention.smithsonian.org/about/about_bio_jerome.aspx.
Page 193
Schmookler, J., *Invention and Economic Growth* (Cambridge, MA: Harvard
    University Press, 1966) 26.
"Jerome Lemelson's Patents," the Lemelson Center for the Study of Invention and
    Innovation, Smithsonian Institution, Washington, DC, invention.smithson-
    ian.org/about/about_patents.aspx.
Page 194
"Lone Wolf of the Sierras," *Design News*, (Cahners Business Information, March
    6, 1995), www.lemelson.org/pdf/lemelson_design_news.pdf.
Page 195
Siegel, R.P., "Down But Not Out," *Mechanical Engineering*, (2004),
    www.memagazine.org/backissues/membersonly/oct04/features/downbut/down
    but.html.
"Jerome Lemelson: An American Inventor," the Lemelson Foundation, www.lemel-
    son.org/innovation/flash/lemelson_book.html.
The Official Lemelson Foundation website, www.lemelson.org/home/index.php.
Page 196
"United States Initiatives," the Official Lemelson Foundation,
    www.lemelson.org/programs/index.php.
Bedi, J., "Jerome Lemelson: Toying with Invention," *Prototype*, (November 2008),
    invention.smithsonian.org/resources/online_articles_detail.aspx?id=529.
Davidson, M., "Jerome Lemelson, Independent Inventor," the Lemelson Center for

the Study of Invention and Innovation, Smithsonian Institution, Washington, DC, invention.smithsonian.org/about/about_bio_jerome.aspx.

Page 197

Brown, K., *Inventors at Work. Interviews with 16 Notable American Inventors* (Microsoft Press), quoted in "Jerome Lemelson: An American Inventor," Lemelson Foundation, 1988, www.lemelson.org/innovation/flash/lemelson_book.html.

Weick, C.W. et al, "Independent Inventors and Innovation: An Empirical Study," *International Journal of Entrepreneurship and Innovation* 6, no. 1 (2005): 5–15.

Lieberman, J. N., *Playfulness: Its Relationship to Imagination and Creativity* (New York: Academic Press, 1977).

Mann, D. "Serious Play," *Teachers College Record* 97, no. 3 (1996): 446–469.

Hutt, C. and R. Bhavnani, "Predictions from Play," *Nature* 237 (1972): 171–172.

Goldmintz, Y. and C.E. Schaefer, "Why Play Matters to Adults," *Psychology and Education: An Interdisciplinary Journal* 44, no. 1 (2007): 12–25.

Brown, Tim, TED Talks, www.ted.com/index.php/talks/tim_brown_on_creativity_and_play.html.

Berg, D., "The Power of Play," *Journal for Quality & Participation* 21, no. 5 (1998): 54–55.

Page 198

National Institute for Play, www.nifplay.org/vision.html.

Bedi, J., "Jerome Lemelson: Toying with Invention," *Prototype*, (November 2008), invention.smithsonian.org/resources/online_articles_detail.aspx?id=529.

Page 199

Lemelson, J. 1953. US Patent Number 2,763,958, filed May 22, 1953, and issued September 25, 1956. p. 2.

Pages 200–201

"Toy Inventor and Designer Guide," Toy Industry Association, www.toyassociation.org/AM/Template.cfm?Section=&Template=/CM/HTMLDisplay.cfm&ContentID=1394.

Mann, D., "Serious Play," *Teachers College Record* 97, no. 3 (1996): 446–469.

Styhre, A., "The Element of Play in Innovation Work: The Case of New Drug Development," *Creativity and Innovation Management* 17, no. 2 (2008): 136–146.

Anderson, J.V., "Creativity and Play: A Systematic Approach to Managing Innovation," *Business Horizons* 37 (1994): 80–85.

Dodgson, M., D. Gann, and A. Salter, *Think, Play, Do: Technology, Innovation, and Organization* (Oxford: Oxford University Press, 2005).

Dougherty, D and C.H. Takacs, "Team Play: Heedful Interrelating as the Boundary for Innovation," *Long Range Planning* 37 (2004): 569–590.

Stanley Mason's Chinese Tallow Tree Plantation

Page 204

Khasru, B.Z., "The Ever Inventive Stanley Mason," *Fairfield County Business Journal* 37, no. 24 (1998): 1–2.

Mason, S.L., "How to Avoid Home Office Distractions," *Morebusiness.com* (April 30, 2001), www.morebusiness.com/getting_started/primer/d987971442.brc.

Page 204

Mason, S.L., "New Invention Ideas: Fill a Need to Be Successful," *Morebusiness.com* (September 4, 2000), www.morebusiness.com/running_your_business/marketing/d967413673.brc.

Mason, S.L., *Inventing Small Products* Crisp Management Library Series (Seattle, WA: Crisp Learning, 1997).

Mason, Stanley, L., "The Chinese Tallow Tree — Growing Oil on Trees," *America's Inventor* (December 1997), www.inventionconvention.com/americasinventor/dec97issue/section12.html.

Bell, M., *Some Notes and Reflection upon a Letter from Benjamin Franklin to Noble Wimberly Jones, October 7, 1772.* (Darien, Georgia: The Ashantilly Press, 1966) 10.

Richman, T. and S. Brenner, "Stanley Mason Is Growing Oil on Trees," *Inc.* (August 1981), www.inc.com/magazine/19810801/oilontrees.html.

Pages 206–207

Richman, T. and S. Brenner, "Stanley Mason Is Growing Oil on Trees," *Inc.* (August 1981), www.inc.com/magazine/19810801/oilontrees.html.

Mason, Stanley, L., "The Chinese Tallow Tree — Growing Oil on Trees," *America's Inventor* (December 1997), www.inventionconvention.com/americasinventor/dec97issue/section12.html.

Interview with John Deere United States offices, December 2008.

Page 208

Webster, C.R., M.A. Jenkins, and S. Jose, "Woody Invaders and the Challenges They Pose to Forest Ecosystems in the Eastern United States," *Journal of Forestry* (October/November 2006): 366–374.

A lag between introduction and establishment and spread of an invasive species is not unusual.

United States Department of Agriculture Plants Database, plants.usda.gov/java/profile?symbol=TRSE6.

The University of Florida Extension, edis.ifas.ufl.edu/AG148.

Mlot, C., "Field Test Backs Model for Invader," *Science* 293 (2001): 1,238.

Page 209

Baldwin, M.J., W.C. Barrow, C. Jeske, and F.C. Rohwer, "Metabolizable Energy in Chinese Tallow Fruit for Yellow-rumped Warblers, Northern Cardinals, and American Robins," *The Wilson Journal of Ornithology* 120, no. 3 (2008): 525–530.

United States Department of Agriculture Plant Guide,
    plants.usda.gov/plantguide/doc/pg_trse6.doc.
Zou, J.W., W.E. Rogers, and E. Siemann, "Increased Competitive Ability and
    Herbivory Tolerance in the Invasive Plant *Sapium sebiferum,*" *Biological
    Invasions* 10, no. 3 (2008): 291–302.

Page 210
Texas Farm Bureau, "Correction August 4 Biodiesel Story, Biofuels Forcing Farm
    Thinking Outside the Box," *Texas Agriculture* (September 1, 2006),
    www.txfb.org/texasAgriculture/2006/090106/090106biodieselcorrection.htm.
Olivier, P.A., "From Field to Factory to Diesel Tank: Empowering Louisiana
    Agriculture," (Self published), www.esrla.com/pdf/tallow.pdf.
Samson, W.D., C.G. Vidrine, and W.D. Robbins, "Chinese Tallow Seed Oil as
    Diesel Fuel Extender," *Transactions of the American Society of Agricultural
    Engineers* 28 (1985): 1,406–1,409.
Crymble, S.D, M.E Zappi,, R. Hernandez, W.T. French, B.S. Baldwin, and T.
    Donald, "Utilization of *Triadica sebifera* as a Novel Biodiesel Feedstock,"
    *American Institute of Chemical Engineers Annual Meeting, Conference
    Proceedings* (New York: American Institute of Chemical Engineers, 2005)
    8,931.
Shupe, T.F., L.H. Groom, T.L. Eberhardt, T.C. Pesacreta, and T.G. Rials, "Selected
    Mechanical and Physical Properties of Chinese Tallow Tree Juvenile Wood,"
    Forest Products Journal 58, no. 4 (2008): 90–93.
Shupe, T.F., L.H. Groom, T.L. Eberhardt, T.G. Rials, C.Y Hse, and T. Pesacreta,
    "Mechanical and Physical Properties of Composite Panels Manufactured from
    Chinese Tallow Tree Furnish," *Forest Products Journal* 56, no. 6 (2006): 64–67.
Texas State Energy Conservation Office, www.seco.cpa.state.tx.us/re_biomass-
    crops.htm.

Page 211
"Biodiesel from Seaweed," *Appropriate Technology* 35, no. 1 (2008): 5.
Pacific Biodiesel, www.biodiesel.com.
Gopalakrishnan, C., K.S. Gadepalli, L.J. Cox, and P. Leung, "The Economics of
    Biomass Energy: A Case Study from Hawaii," *Bioresource Technology* 45
    (1993): 137–143.
Phillips, V.D. "Renewable Resource Development: A Pacific/Asian Perspective,"
    Environmental Science and Technology 24, no. 7 (1990): 958–960.
Hawaii Department of Business, Economic Development and Tourism, *WTTC
    Hawaii Tourism Report 1999.* (London: World Travel and Tourism Council,
    1999), hawaii.gov/dbedt/info/visitor-stats/econ-impact/WTTC99.pdf.
Simberloff, D., "Invasion Biologists and the Biofuels Boom: Cassandras or
    Colleagues?" *Weed Science* 56 (2008): 867–872.
Raghu, S., R.C. Anderson, C.C. Daehler, A.S. Davis, R.N. Wiedgenmann, and D.

Simberloff, "Adding Biofuels to the Invasive Species Fire?" *Science* 313 (2006): 1,742.

Mack, R.N., "Evaluating the Credits and Debits of a Proposed Biofuel Species: Giant Reed *(Arundo donax)*," *Weed Science* 56 (2008): 883–888.

Barney, J.N. and J.M. Ditomaso, "Nonnative Species and Bioenergy: Are We Cultivating the Next Invader?" *BioScience* 58, no. 1 (2008): 64–70.

Page 212

Williamson, M.H., *Biological Invasions* (New York: Chapman Hall, 1996).

Jewkes, J., D. Sawers, and R. Stillerman. *The Sources of Invention* (London: Macmillan and Co. Ltd., 1969).

Buckminster Fuller's Fish-Shaped People Carrier

Page 214

Kenner, H., *Bucky: A Guided Tour of Buckminster Fuller* (New York: William Morrow & Company, Inc., 1973) 47.

Fuller, R.B., *Later Development of My Work*, quoted in J. Meller., ed., *Buckminster Fuller Reader* (London: Jonathon Cape, 1970) 92.

Hall, E.T., *The Hidden Dimension* (1969; New York: Anchor Books Edition, 1990).

Fuller, R.B., *I Seem to Be a Verb* (New York: Bantam Books, 1970).

Marks, R.M., *The Dymaxion World of Buckminster Fuller* (Carbondale, IL: Southern Illinois University Press, 1960) 11.

Fuller, R.B., *Ideas and Integrities* (New York: Collier Books, 1963) 20.

Page 215

Marks, R.W., *The Dymaxion World of Buckminster Fuller* (Carbondale, IL: Southern Illinois University Press, 1960) 16.

Kenner, H., *Bucky: A Guided Tour of Buckminster Fuller* (New York: William Morrow & Company, Inc., 1973) 79–80.

Page 216

The Buckminster Fuller Institute, www.bfi.org/our_programs/who_is_buckminster_fuller/fullers_influence.

The Mitchell and Webb Bronze Orientation Day sketch, www.youtube.com/watch?v=EpeqPdVyQd0.

Page 217

Fuller, R.B., *Ideas and Integrities* (New York: Collier Books, 1963) 24.

Baldwin, J., *BuckyWorks: Buckminster Fuller's Ideas for Today* (Hoboken, NJ: John Wiley & Sons, 1997) 21.

The Buckminster Fuller Institute, www.bfi.org/node/125.
For a list of honorary awards 1952–1983.

Kenner, H., *Bucky: A guided tour of Buckminster Fuller* (New York: William

Morrow & Company, Inc., 1973) 12.

Pages 218–219

Marks, R.W., *The Dymaxion World of Buckminster Fuller* (Carbondale, IL:
Southern Illinois University Press, 1960) 24–25, 34.

"The Dymaxion Bathroom," Buckminster Fuller Institute, 2005–2007,
www.bfi.org/node/548.

Page 222

Fuller, R.B., *Ideas and Integrities* (New York: Collier Books, 1963) 19.

Page 223

Marks, R.W., *The Dymaxion World of Buckminster Fuller* (Carbondale, IL:
Southern Illinois University Press, 1960) 30–31.

Bogomolov, A., "1933 Ford V8 Model 40 and Ford Model 46," Old Timer Gallery
2006, www.autogallery.org.ru/m/ford1933.htm.

The Auto Editors of Consumer Guide, "How Ford Works," How Stuff Works,
auto.howstuffworks.com/ford1.htm

Page 224

For more information on the terrafugia, www.terrafugia.com/vehicle.html.

For more information on the Autovolnator, www.moller.com/.

Email correspondence with Paul Soderman, October 2008.

Fuller, R.B. 1933. US Patent Number 2,101,057, filed October 18, 1933, and issued
December 7, 1937.

Page 225

Marks, R.W., *The Dymaxion World of Buckminster Fuller* (Carbondale, IL:
Southern Illinois University Press, 1960) 31–32.

Pages 226–227

Rossman, J., *The Psychology of the Inventor: A Study of the Patentee* (Washington,
DC: The Inventors Publishing Co., 1931).

Kenner, H., *Bucky: A Guided Tour of Buckminster Fuller* (New York: William
Morrow and Company, Inc., 1973) 68.

Belleflamme, P., "Patents and Incentives to Innovate: Some Theoretical and
Empirical Economic Evidence," *Ethical Perspectives: Journal of the European
Ethics Network* 1 (2006): 267–288.

Leo Szilard and Albert Einstein's Howling Refrigerator

Pages 230–231

Weart, S.R. and G. Weiss Szilard, eds., *Leo Szilard: His Version of the Facts*
(Cambridge, MA: MIT Press, 1978) 6–9.

Krasner-Khait, B. "The Impact of Refrigeration," *History Magazine* (Feb/Mar,
2000), www.history-magazine.com/refrig.html.

Page 232

Junek, O.W., "Methyl Chloride Poisoning from Domestic Refrigerators," *The Journal of the Royal Society for the Promotion of Health*, 51 (1930): 291–292.

Kegel, A.H. and W.D. McNally, "Methyl Chloride Poisoning from Domestic Refrigerators," *Journal of the American Medical Association* 96 (1929): 353–358.

Dannen, G., "The Einstein-Szilard Refrigerators," *Scientific American* (January 1997): 90–95.

This article and Dannen's website (www.dannen.com/szilard.html) are the main sources for the history of the Szilard/Einstein refrigerator, gleaned from interviews with Albert Korodi, an engineer who worked extensively on the prototypes, and other primary sources. Dannen is working on a book about Leo Szilard.

Page 233

Weart, S.R. and G. Weiss Szilard, eds., *Leo Szilard: His Version of the Facts* (Cambridge, MA: MIT Press, 1978) 12, 18.

Pais, A. *"Subtle is the Lord . . . ": The Science and Life of Albert Einstein* (New York: Oxford University Press, 1982).

Telegdi, V.L., "Szilard as Inventor: Accelerators and More," *Physics Today*, October 2000, 25–28.

For more discussion of Szilard's other inventions.

Leo Szilard Online, www.dannen.com/chronbio.html.

Page 234

Weart, S.R. and G. Weiss Szilard, eds., *Leo Szilard: His Version of the Facts* (Cambridge, MA: MIT Press, 1978) vi.

Cooper, R., *Szilard's Ten Commandments*, ASPEC Writer's Workshop, www.eckerd.edu/aspec/writers/Szillard's%2010%20Commendments.htm.

Page 235

Delano, A., "Design Analysis of the Einstein Refrigeration Cycle," (PhD diss., Georgia Institute of Technology, 1998), www.me.gatech.edu/energy/andy_phd/one.htm.

For a more detailed description of how cooling by absorption works.

Page 237

Weart, S.R. and G. Weiss Szilard, eds., *Leo Szilard: His Version of the Facts* (Cambridge, MA: MIT Press, 1978) 12.

Page 238

"Freon®: 1930," Dupont, www2.dupont.com/Heritage/en_US/1930_dupont/1930_freon/1930_freon_ind epth.html.

Page 239

Cole, D.J., E. Browning, and F. Schroeder. *Encyclopedia of modern everyday inventions* (Westport, CT: Greenwood Press, 2003), 218.

"Freon Killed 20 on Russian Submarine," United Press International, November 9, 2008, www.upi.com/Top_News/2008/11/09/Freon_killed_20_on_Russian_submarine/UPI-67371226260351/.

Krezoski, J.R., "Bolton Hall HazMat Incident," University of Wisconsin, Milwaukee, 1997, www.uwm.edu/Dept/EHSRM/CAMPUS/BOL/.

Story of sulfur dioxide leak from an old fridge at the Unversity of Wisconsin, Milwaukee.

Jha, A., "Einstein Fridge Design Can Help Global Cooling," *The Observer*, September 21, 2008, www.guardian.co.uk/science/2008/sep/21/scienceofclimatechange.climatechange

Page 240

Camfridge, www.camfridge.com.

Weart, S.R. and G. Weiss Szilard, eds., *Leo Szilard: His Version of the Facts* (Cambridge, MA: MIT Press, 1978) 14.